RAND NATIONAL DEFENSE RESEARCH INSTITUTE

T0303177

Retention and Promotion of High-Quality Civil Service Workers in the Department of Defense Acquisition Workforce

Christopher Guo, Philip Hall-Partyka, Susan M. Gates

Prepared for the Office of the Secretary of Defense

Approved for public release; distribution unlimited

For more information on this publication, visit www.rand.org/t/rr748

Library of Congress Control Number: 2014959166
ISBN: 978-0-8330-8782-9

Published by the RAND Corporation, Santa Monica, Calif.
© Copyright 2014 RAND Corporation
RAND® is a registered trademark.

Support RAND

Make a tax-deductible charitable contribution at
www.rand.org/giving/contribute

www.rand.org

Preface

The defense acquisition workforce (AW) is responsible for providing a wide range of acquisition, technology, and logistics support (products and services) to the nation's warfighters. The Under Secretary of Defense for Acquisition, Technology, and Logistics (USD [AT&L]) has made it a top priority to support DoD human capital strategies and has directed deployment of a comprehensive workforce analysis capability to support enterprisewide and component assessments of the defense acquisition workforce. The Director, AT&L Human Capital Initiatives, is responsible for departmentwide strategic human capital management for DoD's AW.

This report analyzes data on the civilian AW to examine retention and promotion patterns and their relationship to available measures of workforce quality. It will be of interest to officials responsible for AW planning and management in DoD.

This research was sponsored by USD (AT&L) and conducted within the Forces and Resources Policy Center of the RAND National Defense Research Institute, a federally funded research and development center sponsored by the Office of the Secretary of Defense, the Joint Staff, the Unified Combatant Commands, the Navy, the Marine Corps, the defense agencies, and the defense Intelligence Community. For more information on the RAND Forces and Resources Policy Center, see http://www.rand.org/nsrd/ndri/centers/frp.html or contact the director (contact information is provided on the web page).

Contents

Figures

Tables

Summary

This report examines three topics relevant to civilian employees in the Department of Defense's (DoD's) acquisition workforce (AW). First, we examine available measures of personnel quality and explore whether personnel retention and career advancement vary by quality. Second, we examine the characteristics of workers who rise to the Senior Executive Service (SES) within the AW. This analysis may help us identify leading indicators or aspects of a worker's career that are associated with ascension to the highest levels of DoD. Third, we explore how being in the Civilian Acquisition Workforce Personnel Demonstration Project (AcqDemo) pay plan or other demonstration pay plans affects retention after controlling for workforce quality metrics.

For the first topic, we began with an exploration of potential quality metrics that are readily availability in the administrative data. Our approach to that analysis was based on methods a prior RAND study used to examine these issues with respect to the DoD civilian workforce as a whole (Asch, 2001). While the prior study looked at multiple measures of personnel quality and other personnel outcomes (e.g., pay and promotion) for the entire DoD civilian workforce, our work had a narrower focus. We analyzed data for individuals who entered the DoD civilian acquisition workforce between fiscal years 1998 and 2005. We focused on two measures of personnel quality that are available in the Defense Manpower Data Center data: education and performance ratings. We also explored whether observed relationships between quality and retention are different for different segments of the acquisition workforce, such as services, agencies, and career fields.

One main finding of this analysis was that a higher average performance rating is associated with an increased hazard of separation (decreased retention). Our results also imply that the effect of performance ratings on hazard of separation is much greater for employees who entered the AW at more senior grades: Higher-rated individuals, especially the more senior ones, are more likely to leave DoD, presumably for more-favorable career opportunities. A second finding is with respect to how education, as a measure of personnel quality, affects the hazard of separation. Because many employees upgrade their levels of educations while employed by DoD, we looked at both initial and final educational attainment. Using final education level as the explanatory variable, individuals last observed with a bachelor's, master's, or PhD degree are more likely to be retained than those with less than a bachelor's degree. The intention to upgrade education appears to be an important indicator of retention, but it is not observed clearly. Based on exploratory analysis, our results also suggest that, at least on average, individuals who attain these degrees while in the workforce have lower separation. This finding implies that programs to encourage the AW to pursue higher education while in DoD service may have a strong positive effect on retaining quality workforce.

Regarding the second topic, our analysis of promotion to the SES reveals that organizational characteristics are associated with promotion but that the demographic characteristics of the employee are not. Individuals with an Army background account for almost one-half of the SES ranks, even though they make up just 28 percent of the AW. In contrast, Office of the Secretary of Defense personnel make up 22 percent of the AW but account for less than 10 percent of the SES ranks. Compared to the baseline career field of systems engineering, individuals in production quality, auditing, and program management are more likely to become part of the SES. Individuals in the business career field are less likely to become part of the SES. Gender and race were not significant predictors of whether someone is promoted to the SES, despite the fact that both women and minorities are underrepresented in the AW SES relative to their prevalence in the overall AW. The reason for this discrepancy is that women and minorities tend to work in career fields that are underrepresented in the SES.

Regarding the third topic, our analysis provides evidence that people who enter the AW and were covered by AcqDemo or, in fact, any demonstration pay plan were retained longer compared to those in the General Schedule (GS) plan.[1] Retention increases by 18 percent for the AcqDemo pay plan compared to the GS plan. Entering on another demonstration plan increases retention by 12 percent, and entering on the other pay plans decreases retention by 33 percent compared to the GS plan.

---

[1]   In Appendix C, we find this result holds using percentage of time in one's career spent in AcqDemo as the explanatory variable.

Acknowledgments

The authors would like to thank Beth Asch for her guidance in developing the methodology for this study. We are indebted to Garry Shafovaloff for his ongoing support of and contributions to this work. We are grateful to the file managers at the Defense Manpower Data Center who have provided us with data over the years and answered our many questions about the data files: Marcia Byerly, Portia Sullivan, Leslie Nixon, Michael Kolkowski, Peter Cerussi, and Scott Seggerman. We would also like to thank Edward Keating and Larrie Ferreiro for reviewing the document. Any remaining errors are the full responsibility of the authors.

Abbreviations

AcqDemo	the Civilian Acquisition Workforce Personnel Demonstration Project
AW	acquisition workforce
CF	career field
DAWIA	Defense Acquisition Workforce Improvement Act
DMDC	Defense Manpower Data Center
DoD	Department of Defense
FY	fiscal year
GS	General Schedule
NSPS	National Security Personnel System
OPM	Office of Personnel Management
OSD	Office of the Secretary of Defense
SES	Senior Executive Service
SPRDE	systems planning, research, development, and engineering
USD (AT&L)	Under Secretary of Defense for Acquisition, Technology, and Logistics

CHAPTER ONE

Introduction

The defense acquisition workforce (AW) comprises military personnel, civilian employees of the Department of Defense (DoD), and contractors who perform functions that are related to the acquisition of goods and services for DoD. In 2006, the RAND National Defense Research Institute began a collaboration with DoD to develop data-based tools to support analysis of the organic defense AW, which includes military and DoD civilians, but not contractors.

In response to the Defense Acquisition Workforce Improvement Act (DAWIA) of 1990, DoD has been tracking and reporting on the AW since 1992. The AW is responsible for planning, design, development, testing, contracting, production, introduction, acquisition logistics support, and disposal of systems, equipment, facilities, supplies, or services that are intended for use in, or support of, military missions. A key role of the AW is to oversee the acquisition process. Military and DoD civilian personnel who fulfill one or more of these roles are flagged as being part of the AW. Members of the AW can be found in many different organizations across DoD.

Members of the AW are grouped into career fields. The number and titles of these have changed over time. In fiscal year (FY) 2011, there were 13 main career fields:

- auditing
- business, cost estimating, and financial management[1]
- contracting
- communications and information technology
- facilities engineering
- industrial property management
- life-cycle logistics
- program management oversight and program management
- purchasing and procurement
- quality assurance
- science and technology
- systems planning, research, development, and engineering (SPRDE)[2]
- test and evaluation engineering.

---

[1] The business career field comprises two career paths, cost estimating and financial management.

[2] The SPRDE workforce currently comprises two separate career fields: SPRDE–Systems Engineering and SPRDE–Program Systems Engineer. The former career field is roughly 100 times larger than the latter one. In our analysis, we combine these two career fields into a single SPRDE career field.

As Gates et al., 2013, describes, RAND has assembled a comprehensive data file that can support a DoD-wide analysis of DoD AW. The RAND data file comprises information drawn from several files that the Defense Manpower Data Center (DMDC) maintains, including the DoD civilian personnel inventory file and AW person and position files.

In the DMDC database, records can be linked across files in useful ways. By linking records across time and across files, we were able to examine movement into and out of the AW, movement between the DoD military and civilian workforces, and promotion and experience trajectories.

Our prior analyses have provided a descriptive analysis of the AW and its retention patterns. The analyses on retention did not consider workforce quality. It is well known that turnover is not always a bad thing for an organization. Defense workforce managers strive to understand the quality of those who are retained relative to that of those who separate. In particular, they want to retain high-quality personnel. One strategy that managers across DoD have employed to improve workforce quality is the implementation of personnel demonstration projects. The 1978 Civil Service Reform Act authorized the Office of Personnel Management (OPM) to approve a limited number of demonstration programs to test improved personnel management procedures.[3] Through these demonstration programs, federal agencies are allowed to set up alternatives to the General Schedule (GS) pay system that governs most federal white-collar employees. The objective of allowing OPM to waive existing laws and regulations governing human resources practices in Title 5 of the U.S. Code is "to propose, develop, test, and evaluate alternative approaches to managing its human capital" (U.S. General Accounting Office, 2004, p. 7).

DoD laboratories were among the early adopters of these human capital reform efforts. The first demonstration project OPM approved, the "China Lake" demonstration project, was implemented by the Navy at the Naval Weapons Center in China Lake, California, and the Naval Ocean Systems Center in San Diego, California. The perceived success of the China Lake demonstration project led Congress to authorize OPM to sustain and expand such efforts. The National Defense Authorization Act for Fiscal Year 1995, Section 342, made permanent the China Lake demonstration and authorized the Secretary of Defense to establish the DoD Science and Technology Reinvention Laboratory Demonstration Program and removed any mandatory expiration date for the laboratory's demonstration projects. Demonstration projects were instituted among employees at the Naval Research Laboratory, Naval Sea Systems Command Warfare Center, Air Force Research Laboratory, and several Army research laboratories. Many of these organizations employ members of the AW.

The National Defense Authorization Act for FY 1996 authorized OPM to develop a program targeting the members of the DoD AW who were not already covered by other demonstration projects. The Civilian Acquisition Workforce Personnel Demonstration Project (AcqDemo) was implemented in 1999 and grew to cover 11,416 employees by September 2006 (Werber et al., 2012).[4] The objective of AcqDemo was to create a civilian personnel system that effectively supported the DoD acquisition mission.

---

[3] Title VI of the Civil Service Reform Act is now codified in 5 U.S.C. 4703, Demonstration Projects.

[4] In 2007, DoD transferred most AcqDemo employees into the National Security Personnel System (NSPS), which embodied key demonstration project characteristics and was intended to be the primary personnel system for all DoD civilian employees. However, NSPS was dissolved in 2011, and employees were transitioned back into AcqDemo. Between 2007

This report examines some available measures of the quality of the DoD AW and explores whether personnel retention and career advancement vary by quality. We also explore whether observed relationships between quality and retention are different for different segments of the AW, such as services, agencies, career fields, and pay plans. We begin with an exploration of potential quality metrics that are readily availability in the administrative data. Our approach to the analysis is based on methods a prior RAND study (Asch, 2001) used to examine these issues with respect to the DoD civilian workforce as a whole. While the prior study looked at multiple measures of personnel quality and other personnel outcomes (e.g., pay and promotion) for the entire DoD civilian workforce, our work takes a narrower focus. We analyze data for individuals who entered the DoD civilian acquisition workforce between FYs 1998 and 2005. We focus on two measures of personnel quality that are available in the DMDC data: education and performance ratings. We then examine the characteristics of workers who rise to the senior executive service within the acquisition workforce. This analysis may help us identify leading indicators or aspects of a worker's career that are associated with ascension to the highest levels of DoD. Finally, we explore how being in AcqDemo or other demonstration pay plans affects retention after controlling for workforce quality metrics.

---

and 2011, about 3,000 AW employees, primarily unionized Army civilians, did remain in AcqDemo. (See Werber et al., 2012.)

CHAPTER TWO

Retention of High-Quality Civil Service Workers in the Acquisition Workforce

Approach

The intent of this analysis was to explore whether DoD retains higher-quality DoD AW civilian personnel at higher rates than it does lower-quality DoD AW civilian personnel. In the process, we also gained insight into other workforce characteristics that are associated with retention. This chapter begins by describing candidate measures of personnel quality, their determinants, and how they may be correlated with retention. We then explain our outcome metric for retention—time to separation.[1]

Personnel quality, which we define as an individual's productivity at a job, depends on innate ability, motivation, and job-specific factors that determine whether the individual is a good match. Although data on these general determinants of quality are scarce, the DoD civilian master file contains several variables that might correlate with quality. One candidate quality measure is the level of education on entry into the workforce. An advantage of using education as a measure of quality is that the entry education level is easily observable. However, although education captures an individual's general skill level—applicable to both civil service and other job opportunities—education does not capture the individual's fit with a particular civil service job (Asch, 2001). Performance ratings may be better indicators of an individual's fit with a job but can suffer from measurement error and bias. The rating of an individual is determined by a supervisor, and in addition to worker quality, other factors that may determine the performance rating include the methods used to monitor the worker's output, the frequency with which the supervisor is able to monitor output, the cost of monitoring output, and the supervisor's own subjective bias.[2] In the section on data, we will provide a detailed description of the acquisition workforce rating system.

In the next section, we will discuss survival analysis, specifically the Cox proportional-hazard model. This model is well suited to studying the occurrence and timing of separation from censored data.[3] The outcome of interest in this analysis is the length of stay until separation from the AW. To determine what explanatory variables to include in the Cox regression

---

[1] This analysis sometimes uses alternative terms, such as *rate of separation*, *probability of retention*, or *hazard of separation*, but all refer to the same effect.

[2] Asch, 2001, also identifies promotion speed as a measure of quality, and our preliminary analysis did find that time to promotion is highly correlated with performance rating. We therefore chose to focus our analysis on performance rating.

[3] Other conventional regression models focus on only the occurrence of separation but are insensitive to the timing of the event. For these models, all that matters is the binary outcome (separation or no separation), so a separation after one year is treated the same as a separation after 20 years. In addition, these other models often struggle with "censored" data. When

and to verify necessary model assumptions, we conducted univariate analysis of the effects of individual predictors on survival. This analysis can be found in Appendix A.

Survival Analysis Model

In survival analysis, the *hazard function* describes the probability distribution of the time to separation. It is interpreted as the probability that an individual experiences the event of interest (separation) at a particular time given that the event has not yet occurred. More formally, the hazard function, $h(t)$, is expressed as the probability-density function, $f(t)$, divided by the cumulative survival function, $S(t)$:

$$h(t) = \frac{f(t)}{S(t)},$$

$$\text{where } S(t) = \Pr\{T > t\} \text{ and } f(t) = \frac{dS(t)}{dt}.$$

The cumulative survival function, $S(t)$, measures the cumulative probability that the time of employee separation, T, occurs after month t. The hazard function, $h(t)$, measures the probability of separation occurring at month t, given that it did not occur at month $t-1$.

In the Cox-proportional-hazard model, the hazard function is given by:

$$h_i(t) = h_0(t)\exp\left(\beta^T x_i\right),$$

where $h_i(t)$ is the hazard of separation for individual i; $h_0(t)$ is the baseline hazard function; β is a vector of regression coefficients; and x_i is the vector of explanatory variables. The baseline hazard is the probability of separation when every covariate in x_i is zero, and this function is left unspecified—i.e., the model does not estimate an *absolute* level of separation risk for the hazard function, $h_i(t)$. Instead, exponentiated regression coefficients are interpreted as *relative* hazard ratios between one set of explanatory variables for individual i and another set for individual j:

$$\frac{h_i(t)}{h_j(t)} = \frac{\exp\left(\beta^T x_i\right)}{\exp\left(\beta^T x_j\right)} = \exp\left[\beta^T\left(x_i - x_j\right)\right].$$

The main assumption here is that the hazard ratio is constant across time. The vector of regression coefficients, β, is estimated using partial-likelihood techniques, and each coefficient estimate indicates the effect of the corresponding explanatory variable on the hazard of separation (i.e., the monthly separation rate), holding all other covariates constant. A positive coefficient signifies that the explanatory variable increases the hazard of separation and reduces retention. Explanatory variables in x_i include measures of personnel quality (education and average performance rating) and other job and individual characteristics measured on entry that do not vary with time.

---

an observation is censored, the subject did not experience the event of interest (separation) during the time the subject was part of the study, although the individual is expected to separate from the workforce eventually.

Data Description

Three databases were combined to create a longitudinal data set of personnel in the acquisition workforce. DMDC provided DoD civilian personnel inventory and transactions files. The inventory data file captured personnel characteristics of all DoD civilians at the end of the fiscal year, dating back to 1980. The transaction data file captured any personnel changes that occurred during a year (e.g., appointments, reappointments, promotions, and separations) and helped us determine the time between entry and separation for each individual. By using the exact dates of entry and exit, we could be more precise in the time spent in employment than using just the year would allow. Finally, the DAWIA acquisition workforce person file, which dates back to 1992, helped us determine which DoD employees were in the AW.

The constructed longitudinal data set we used tracks, through FY 2012, the careers of eight cohorts of civilian personnel who entered the AW workforce between FY 1998 and FY 2005. A cohort is defined as those who entered the AW either from elsewhere in DoD or from outside DoD in a certain fiscal year. The analysis focuses on the careers of those who entered after 1998 because the performance ratings from before 1998 are incomplete. Individuals who entered after 2005 are not included, so that at least several years of employment data are available for the newest cohort.

Looking only at the AW subset within the larger DoD civilian workforce introduced issues of how to define an entry or separation, particularly in the case of transfers, when individuals move between the AW and another part of DoD. Between 2001 and 2004, DoD undertook a major effort to rationalize (make consistent) the definition of AW, resulting in numerous administrative transfers. A transfer is defined as *administrative* if the agency, bureau, occupational series, functional occupational group, and pay plan remain the same. In contrast, the transfer is considered *substantive* if any category changes (Gates et al., 2008). For our purposes, an *entry* is defined as someone who is either new or who was previously employed by DoD and made a substantive transfer into the AW. Administrative transfers into the AW are not considered as new entries in a given cohort year.

Information attached to an individual's career profile includes job and individual characteristics measured on entry and how these characteristics change over time. Job characteristics recorded include component (Army, Navy or Marines, Air Force, or other defense agency), occupational area, position held, months of federal service, pay, grade, performance ratings, and time of separation (if observed). To include non-GS employees, standardized grade categories (e.g., low, middle, and senior) were created across different pay plans.[4] Individual characteristics include gender, race and ethnicity, age, education, geographic region, veteran status, and handicap status. Table 2.1 provides the variable names and means of the characteristics included in our longitudinal data set for cohorts of new entrants to the AW.

Some differences across cohorts reflect changes in the type of positions that make up the AW, as well as demographic shifts in the workforce. The percentage of females in each cohort fell from 43 percent in 1998 to 35 percent in 2005. Over this period, the level of experience of

---

[4] Entry level includes GS (and related) 1–8; YA, YB, YP, YD, YE, YH, YI, YK, YL, YM, DR 1; NM 2; DA, DB, DE, DJ, DP, ND, NH, NJ, NK, NO, NP,NR, NT 1–2; DK 1–3.

   Midlevel includes GS (and related) 9–13; YC, YF, YJ, YN-1; YA, YB, YD, YE, YH, YI, YK, YL, YM, DR-2; YB, YE, YL, DB, DE, DJ, DP, NH, NK, NO, NP, NM-3; DK-4; DA, ND, NJ, NR, NT 3–4; IA-2–3.

   Senior level includes GS (and related) 14-15; YC, YF, YJ, YN 2 and 3; YA, YD, YH, YI, YK, YM 3; YB, YE, YL, DB, NH, NP 4; DE, DJ, IA, NO, NM 4–5, DR 3–4; DP, NR, ND 5; DA, NT 5–6; IP (all).

Table 2.1
Variable Means, FY 1998–2005 Cohorts

Variable	FY								
	1998	1999	2000	2001	2002	2003	2004	2005	Overall
Average age	38.8	37.2	36.2	41.7	36.6	37.2	36.3	36.6	37.7
Average months of service	127.4	102.1	88.1	152.2	80.0	80.4	74.4	65.4	94.8
Average compensation ($000)	40.6	40.7	42.5	53.2	50.1	46.1	50.2	49.1	48.7
% who are									
Supervisors	7.3	4.9	6.7	12.4	4.8	5.0	5.6	4.1	6.6
Veterans	25.6	25.4	22.4	24.8	27.5	29.1	24.0	32.1	26.8
Female	43.3	43.4	40.9	34.3	28.7	34.7	31.7	35.3	34.4
Not white or Hispanic	24.9	23.8	27.5	23.0	24.1	23.5	21.8	23.7	23.7
Handicapped	8.0	5.6	7.9	9.0	8.2	7.8	7.2	7.5	7.9
Grade									
Entry (%)	39.8	45.5	49.0	21.8	33.3	40.2	40.0	36.4	35.4
Middle (%)	56.8	51.9	47.0	69.9	62.3	56.1	55.5	60.2	59.8
Senior (%)	3.4	2.6	3.9	8.3	4.4	3.8	4.5	3.4	4.8
Substantive transfer in (%)	50.0	47.4	43.7	42.9	42.0	43.9	48.2	47.8	45.2
Performance rating									
Average	3.8	3.9	3.7	4.2	3.8	4.5	4.2	4.2	4.1
% with a rating of									
1	0.0	0.0	0.0	0.0	0.0	0.0	0.0	0.0	0.0
2	0.0	0.0	0.2	0.0	0.0	0.2	0.0	0.0	0.1
3	40.3	34.9	43.5	22.9	40.6	8.2	24.5	20.1	28.5
4	33.3	32.5	33.8	27.9	25.4	29.7	21.4	30.2	27.8
5	24.4	28.8	17.2	46.3	26.9	59.4	46.9	40.1	37.9
Organization (%)									
Army	27.2	27.2	16.4	59.6	34.3	89.5	66.0	55.1	49.9
Navy and Marines	51.5	41.1	53.2	32.6	53.6	6.0	33.4	28.6	37.5

Table 2.1—Continued

Variable	FY								Overall
	1998	1999	2000	2001	2002	2003	2004	2005	
Air Force	0.7	24.6	16.0	7.3	11.9	4.3	0.6	0.6	7.0
OSD or other	20.6	7.1	14.4	0.5	0.2	0.2	0.1	15.8	5.5
Education (%)									
No college	15.3	15.7	11.7	18.2	14.1	14.7	13.9	14.6	15.0
Some college	15.5	12.7	7.9	10.3	7.5	11.5	9.8	10.8	10.1
Bachelor's degree	45.1	48.8	59.8	46.0	55.6	48.2	54.1	52.0	51.7
Master's degree	17.7	17.5	16.0	18.8	17.8	19.1	17.1	17.7	17.8
PhD	2.0	2.3	1.5	3.6	2.3	2.4	2.1	1.4	2.3
Career field (%)									
Systems engineering	19.0	21.6	31.8	32.7	41.5	27.2	31.9	22.3	19.0
Program management	4.6	5.8	4.7	10.6	6.0	9.9	6.7	6.6	4.6
Contracting	19.9	25.1	19.3	10.8	13.5	13.0	13.0	14.6	19.9
Purchasing	7.9	4.3	2.1	1.1	1.5	0.6	1.1	1.3	7.9
Production quality	9.4	0.7	1.5	1.8	1.6	5.6	5.0	6.9	9.4
Business	9.3	9.9	8.4	11.2	5.4	10.5	6.1	7.1	9.3
Life-cycle logistics	11.5	13.2	10.7	13.3	9.4	11.3	13.3	18.5	11.5
Information technology	4.4	3.4	1.2	5.4	5.5	6.4	2.5	3.2	4.4
Auditing	4.7	6.6	4.1	4.5	8.2	8.4	6.0	3.9	4.7
Facilities engineering	0.1	3.2	13.0	0.0	0.0	0.0	0.0	10.6	0.1
Test evaluation	0.0	0.0	0.0	1.6	4.2	2.9	11.8	2.8	0.0
Missing or other	9.2	6.3	3.2	7.0	3.1	4.1	2.6	2.4	9.2
Individuals	1,183	1,340	1,711	4,471	4,770	1,568	3,867	4,154	23,064

new entrants generally declined—the average months employed by DoD at the time of entry into the AW fell from 127 to 65 months. However, this trend did not hold for the FY 2001 cohort, which saw an abrupt increase in very experienced, middle- and senior-grade hires. The FY 2003 cohort saw a high percentage of new hires into the Army. These civilian hires were necessary to replace military service members who had been transferred into the infantry at the time.

Figure 2.1 shows how the cohort size and composition, in terms of career fields, has changed over the years.[5] Each bar represents that year's cohort, and the vertical height of a colored band represents the number of individuals within the corresponding career field. Most noticeably, the systems engineering career field has grown significantly. Despite these differences, each cohort had approximately similar proportions of entrants by ethnicity, handicap status, and transfer status over the eight-cohort study period.

Quality Metric 1: Education

The first measure of personnel quality we apply is education. Our prior work (Asch, 2001) used only the education level *on entry* in the Cox regression because of a concern that the education variable was not accurately and consistently updated in the civilian data files. Table 2.1 displays educational attainment of the workforce on entry into the AW for our cohort sample. Table 2.2 describes the percentage of workers in each cohort for whom we observed change in educational attainment during their careers in the AW. Fifteen to 25 percent of individuals

Figure 2.1
Individuals in Each Career Field, FY 1998–2005 Cohorts

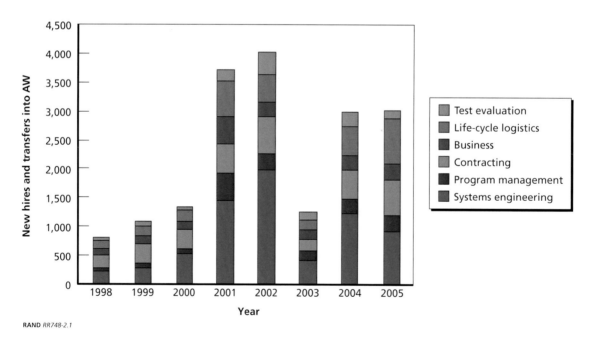

---

[5] Some individuals enter the AW with a career field value of "Z" for missing and are then assigned a career field during their second year of employment. In most years, less than 5 percent of entries fall into this category, but in 2001, 42 percent of entries into the AW were coded without a career field. To account for this, we instead used the career field from the second year in constructing career field data for all cohorts. The initial career field is used in cases where an individual worked for the AW for just one year or has a missing value in the second year and a nonmissing value in the first year.

within each cohort of new AW civilian personnel attained education upgrades while they were in the AW, with most upgrades consisting of master's degrees.

Table 2.3 provides a glimpse at the number of individuals in our data set making educational transitions from starting education (observed on entry) to final education (either observed on leaving the workforce or in 2012, if they were still in the workforce). The yellow diagonal boxes signify the numbers of individuals who did not change their education level. Below the diagonal, in the lower-left half, the education levels of small numbers of individuals fell, most likely due to errors in recording the entry education level. The cells above the diagonal, in the upper-right half, show the number of individuals who increased their education levels during the observation period. For example, 3,038 (2,487 + 226 + 325) individuals obtained master's degrees while they were employed in the AW, as compared to 3,992 individuals who entered with a master's degree in hand. Over 40 percent of individuals in these cohorts who had master's degrees as of 2012 had obtained their degrees while employed in the AW. Likewise, over 30 percent of individuals in these cohorts who had attained doctorates by 2012 had upgraded to this education level while in the workforce. In the Results section of this chapter, we will discuss the implications of educational upgrades for using education as a quality metric.

Quality Metric 2: Performance Rating

The second measure of quality is the average performance rating over the period an individual is in the data set. Certain caveats attributed to data quality should be noted. First, we include only personnel on the "H" performance rating plan, which has a five-step scale, with 1 being

Table 2.2
Percentage of Individuals with an Education Upgrade, FY 1998–2005 Cohorts

	Year							
	1998	1999	2000	2001	2002	2003	2004	2005
New educational level								
Bachelor's	5.3	5.3	5.5	3.5	5.0	6.3	3.7	5.9
Master's	10.4	17.6	17.2	10.3	14.7	14.0	12.9	13.5
PhD	1.3	2.1	1.0	0.8	1.0	0.8	1.0	0.7
Total	17.0	25.0	23.7	14.6	20.8	21.1	17.6	20.0

Table 2.3
Starting and Final Education Levels, FY 1998–2005 Cohorts

Starting Education	Final Education				
	No College	Some College	Bachelor's	Master's	PhD
No college	2,468	194	436	325	32
Some college	19	1,541	523	226	9
Bachelor's	7	16	9,337	2,487	79
Master's	5	0	11	3,992	99
PhD	6	0	13	16	492

the lowest.[6] As Table 2.1 shows, it is rare in practice for someone to receive a 1 (unsatisfactory) or a 2 (minimally successful), but there is an even distribution of ratings between 3 (fully successful), 4 (exceeds fully successful), and 5 (outstanding). Employees can also receive an X if not rated in a given year, which generally occurs for new employees in their first year. Of our FY 1998–2005 cohorts, 78 percent of the civilian employees in the AW were on the "H" performance plan; 7,411 were dropped because they were on performance plans other than "H" for any year. The other performance plans were either three-step or two-step (pass or fail) scales and suffered from limited variation (e.g., almost all individuals on the pass or fail performance plan passed). With no way to compare ratings on a two- or three-step scale with ratings on a five-step scale, we focused only on individuals in the "H" plan, for which the most useful performance ratings exist. A possible issue arises if various workers are systematically selected into different performance rating plans. We assumed what determines a worker's performance plan is independent of their propensity to stay within the AW. Second, employees generally do not receive performance ratings their first year in the civilian DoD AW. If these individuals separated from the AW before receiving a rating, they were excluded from our longitudinal data. In fact, 1,611 individuals were dropped for not having a rating. If the employees who were dropped from our data set are different from employees who stayed long enough to receive a rating, our analysis would not fully capture the reasons the dropped individuals separated. Finally, while individual performance ratings change over time, this analysis uses the average rating as an indication of overall quality. Supervisor ratings given to an employee year to year may vary in response to idiosyncrasies or outside circumstances. For example, a supervisor may inflate a rating in a certain year if he or she perceives a need for an outstanding rating to advance an imminent promotion or may deflate a rating if the "perceived" need for a top rating is diminished. Variation over time in performance ratings is less of an issue if we assume these idiosyncrasies are random each year. However, using the average rating overlooks trends in an employee's performance—for example, in the case of an employee beginning with low ratings but consistently improving every year.

As Table 2.1 shows, the average performance rating fluctuates across cohorts, with a general trend toward higher average ratings for each cohort. From 1998 to 2005, the average rating by cohort increased from 3.8 to 4.2. While the proportion of 4s held constant, the instance of 3s decreased, and 5s increased by similar magnitudes. Table 2.4 shows that the distribution of performance ratings also varies by service.[7]

Dependent Variable: Time to Separation

The dependent variable, time to separation from the AW, is denominated by month. It was created by subtracting the date of entry from the inventory file and the date of separation from the transaction file.[8] In cases of censored data (i.e., separation is not observed before the data ends

---

[6] During the period studied in Asch, 2001, 1 was the best rating and 5 was the worst rating. The scale switched directions in 1997.

[7] We acknowledge the noticeable differences in performance ratings between services but cannot speculate on why, for example, the Army shows a much higher percentage of 5s than the other services do. To account for the potential impact of these performance rating differences on retention rates, we also ran a preliminary specification that allowed the effect of a higher rating to vary depending on both grade and service. In general, the marginal effects of performance rating on retention for each grade and service combination were consistent with the results we chose to include.

[8] We dropped 2,564 observations from our sample because the transaction file did not record both when they entered the AW and when they left the AW, preventing us from determining the time to separation.

Table 2.4
Distribution of Performance Ratings, by Service, FY 1998–2005 Cohorts

Performance Rating	Percentage				
	Army	Navy	Marines	Air Force	Other/OSD
1	0.0	0.1	0.0	0.0	0.0
2	0.0	0.2	0.0	0.2	0.0
3	3.2	69.9	45.4	31.2	30.1
4	25.4	26.9	44.5	47.9	57.0
5	71.3	2.9	7.9	20.4	12.5

in FY 2012), time to separation was set to the number of months until the data ends. As mentioned previously, classification of a transfer as a separation requires distinguishing between administrative and substantive transfers. For our purposes, a separation occurs when someone leaves DoD entirely or makes a substantive transfer out of the AW to elsewhere in DoD. Individuals making administrative transfers within the AW are counted as staying within the workforce, and individuals making administrative transfer out of the AW are excluded from the analysis.

Results

This section first presents overall results on the effect of education and performance rating on retention. Two main specifications of the Cox regression model were estimated for the FY 1998–2005 cohorts. While both include average performance rating, they differ in that specification (1) uses education on entry and specification (2) uses final observed education. As mentioned in the data description, we controlled for a wide range of relevant observable variables, such as job and individual characteristics. Of special interest are the columns labeled "Hazard Ratio." For continuous variables, such as average performance rating, the hazard ratio, which is equal to the exponential of the coefficient estimate, gives the estimated percentage change in the hazard for a one unit increase in the covariate. *A hazard ratio greater than 1 implies an increase in the hazard of separation (decrease in retention), and a ratio less than 1 implies a decrease in the hazard of separation (increase in retention).* For example in specification (1) of Table 2.5, the estimated hazard ratio for performance rating is 1.367. A one-unit increase in average performance rating increases the hazard of separation by 36.7 percent (1.367 − 1). The hazard of separation is the probability that an employee separates at a given time, given that he or she has not separated before then.

For indicator and dummy variables, the hazard ratio can be interpreted as the ratio of the estimated hazard for those with a value of 1 to the estimated hazard for those with a value of 0, holding all other covariates equal. For example, the estimated hazard ratio for "Supervisor" in specification (1) means the hazard of separation is 33.4 percent higher for those with supervisor status than for those without (i.e., probability of retention decreases by 33.4 percent).

Table 2.5
Cox Regression Model of Months to Separation, FY 1998–2005 Cohorts—Specifications (1) and (2)

Variables	Time to Separation					
	Specification (1)			Specification (2)		
	Estimate	Std. Error	Hazard Ratio[a]	Estimate	Std. Error	Hazard Ratio[a]
Performance rating	0.312***	0.030	1.367***	0.300***	0.030	1.349***
Average age	0.020***	0.002	1.020***	0.019***	0.002	1.019***
Average months of service	0.004***	0.000	1.004***	0.004***	0.000	1.004***
Compensation ($000)	−0.010***	0.001	0.990***	−0.007***	0.001	0.993***
Those who are						
Supervisors	0.288***	0.053	1.334***	0.307***	0.053	1.359***
Veterans	−0.172***	0.036	0.842***	−0.163***	0.036	0.850***
Female	−0.016	0.035	0.984	−0.026	0.035	0.974
Not white or Hispanic	−0.020	0.033	0.980	−0.010	0.033	0.990
Handicapped	0.082*	0.049	1.085*	0.084*	0.049	1.087*
Organization						
Army	0.196**	0.091	1.216**	0.220**	0.091	1.246**
Navy or Marines	−0.066	0.090	0.936	−0.084	0.090	0.920
Air Force	−0.168	0.106	0.846	−0.068	0.106	0.935
Grade						
Middle	−0.109**	0.043	0.896**	−0.139***	0.043	0.870***
Senior	0.008	0.086	1.009	0.017	0.087	1.017
Substantive transfer in	−0.557***	0.032	0.573***	−0.561***	0.032	0.571***
Initial education level						
Bachelor's	−0.148***	0.036	0.863***			
Master's	−0.143***	0.045	0.867***			
PhD	0.140	0.089	1.151			
Final education level						
Bachelor's				−0.305***	0.038	0.737***
Master's				−0.654***	0.044	0.520***
PhD				−0.328***	0.087	0.720***
Career fields						
Program management	0.598***	0.056	1.819***	0.545***	0.056	1.724***
Contracting	0.095*	0.056	1.100*	0.133**	0.056	1.143**
Purchasing	0.367***	0.098	1.444***	0.284***	0.098	1.329***
Production quality	0.231***	0.075	1.260***	0.138*	0.076	1.148*
Business	0.407***	0.057	1.502***	0.349***	0.058	1.418***
Life-cycle logistics	−0.292***	0.060	0.746***	−0.371***	0.060	0.690***
Information tech	0.762***	0.064	2.142***	0.645***	0.064	1.907***
Test evaluation	−0.076	0.084	0.927	−0.080	0.084	0.923
Auditing	0.803***	0.128	2.233***	0.837***	0.128	2.309***
Facilities engineering	0.825***	0.073	2.283***	0.813***	0.073	2.255***
Missing or left out CF	0.724***	0.063	2.062***	0.668***	0.063	1.950***

Table 2.5—Continued

Variables	Specification (1)			Specification (2)		
	Estimate	Std. Error	Hazard Ratio[a]	Estimate	Std. Error	Hazard Ratio[a]
Year dummies						
1999	−0.222***	0.075	0.801***	−0.227***	0.075	0.797***
2000	−0.526***	0.081	0.591***	−0.541***	0.082	0.582***
2001	−0.010	0.060	0.990	−0.040	0.060	0.960
2002	−0.413***	0.065	0.661***	−0.446***	0.065	0.640***
2003	−0.344***	0.075	0.709***	−0.359***	0.075	0.698***
2004	−0.246***	0.068	0.782***	−0.299***	0.068	0.741***
2005	−0.478***	0.069	0.620***	−0.518***	0.069	0.596***
Censored observations	6,139	6,139	6,139	6,139	6,139	6,139
Observations	21,714	21,714	21,714	21,714	21,714	21,714

*** 1-percent significance, ** 5-percent significance, * 10-percent significance.

[a] Hazard Ratio = exp(Estimate)

Education

We grouped individuals into four educational attainment categories: "less than bachelor's," "bachelor's," "master's," and "PhD."[9] Because "less than bachelor's" is the omitted category in the regression, the other three education categories are compared relative to the baseline education category consisting of educational attainment short of a bachelor's degree (no college, some college, and associate's degree). The hazard ratio associated with "bachelor's" in specification (1) is 0.863, which is less than 1. This means the hazard of separation is 13.7 percent (1 − 0.863) lower for those with a bachelor's degree than for those without a college degree (i.e., the probability of retention is 13.7 percent greater). In specification (2), the education indicator variables establish the education levels when individuals leave or when last observed.

How we define education (either education on entry or final observed) affects the estimates of the probability of retention. When using education on entry as a predictor of retention, as in (1), we found that the probability of retention is greater for those with bachelor's or master's degrees than for those without college degrees. The findings from Asch, 2001, also used initial education level as the explanatory variable. That work analyzed the FY 1988 and FY 1992 cohorts of the entire civilian DoD workforce, as opposed to our analysis of the FY 1998–FY 2005 civilian AW cohorts. Our eight-cohort sample shows a greater frequency of advanced degrees on entry into the AW than the FY 1988 and FY 1992 DoD cohorts did. Our finding is in line with the prior study, which found the same direction of effect from having a bachelor's degree, although it was not as strong—Asch, 2001, p. 59, found a 7-percent reduction in separation hazard for the FY 1992 cohort, while our analysis found a 14-percent reduction (depending on whether initial or final education level is used in the regression). However, Asch, 2001, did not find a significant effect from having a master's degree, while our analysis of

---

[9] Over the last 20 years, there has been a trend toward hiring more-educated personnel into the AW. Consequently, the group of those who did not attain a college degree (i.e., no college, some college, and associate's degree) is relatively small, and we combined these into a single education category.

the AW found a significant 13-percent reduction in separation hazard from having a master's degree. Like Asch, 2001, we also found no significant effect on retention from entering with a PhD, when using initial education level as the explanatory variable.

When using final observed education in (2), we find a positive effect on retention for all education levels. Leaving with a bachelor's, master's, or PhD implies a higher probability of retention (26 to 48 percent higher) than leaving with an education level below a bachelor's degree does. The magnitude of retention probability is also greater moving from (1) to (2). In specification (2), we included both those who entered with the degree and those who upgraded to the degree while in the workforce. One hypothesis is that those who upgraded might have a higher probability of retention than those who entered with a PhD, so including the upgraders increases the probability of retention, which could explain the differences between (1) and (2).

We can see that the populations of nonupgraders and upgraders exhibit observable differences. The characteristics of nonupgraders who entered the AW with that level of education are presented on the left side of Table 2.6. The characteristics of educational upgraders are presented on the right side of the table. Across every education category, nonupgraders were older, had more months of experience, were more likely to be supervisors, had a higher average salary, and were less likely to be female.

In exploratory analysis, we observed that those who upgraded to a bachelor's, master's, or PhD in a certain year tended to stay longer after upgrading than new hires who had entered during that same year. For each year, we compared the time until separation for those who

Table 2.6
Summary of Acquisition Workforce, by Education Upgrade, FY 1998–2005 Cohorts

	Final Education						
	No Education Upgrade				Education Upgrade		
	No College	Bachelor's	Master's	PhD	Bachelor's	Master's	PhD
Personnel	2,487	9,337	3,992	492	959	2,487	99
Average age	45.1	34.4	40.9	44.4	34.7	31.5	39.6
Average months of service	178.2	70.4	89.2	97.9	82.4	45.7	62.9
Average compensation ($000)	46.9	47.4	57.3	67.1	41.5	42.7	57.5
% who are							
Supervisors	5.80	5.10	12.80	18.10	3.2	2.40	7.10
Female	46.00	30.20	24.80	16.50	42.7	38.80	21.20
Veterans	38.80	18.80	35.70	18.50	27.7	19.90	31.30
Minorities (%)	22.20	24.50	21.00	20.10	26.4	25.30	24.20
Organization (%)							
Army	61.40	45.40	49.90	48.00	54.4	44.50	38.40
Navy or Marines	31.80	44.00	35.40	36.20	31.3	36.10	43.40
Air Force	3.00	4.10	9.30	14.40	9.3	13.30	17.20
OSD or other	3.90	6.50	5.40	1.40	5.0	6.00	1.00
Grade (%)							
Entry	24.70	42.70	17.10	6.50	51.3	53.70	17.20
Middle	72.70	53.90	71.60	70.30	46.9	44.50	75.80
Senior	2.60	3.40	11.30	23.20	1.6	1.80	7.10

Table 2.6—Continued

| | Final Education | | | | | | |
| | No Education Upgrade | | | | Education Upgrade | | |
	No College	Bachelor's	Master's	PhD	Bachelor's	Master's	PhD
Career field (%)							
Systems engineering	2.20	38.10	32.50	51.40	23.4	35.70	69.70
Program management	8.70	3.60	8.50	7.10	4.3	4.00	1.00
Contracting	4.40	13.70	13.90	6.50	17.2	20.40	6.10
Purchasing	5.70	0.60	0.50	0.00	6.9	0.50	0.00
Production quality	8.00	2.30	1.90	0.80	3.6	2.70	0.00
Business	9.90	4.80	5.10	1.20	5.7	4.70	4.00
Life-cycle logistics	21.00	7.80	7.20	3.00	12.9	9.20	6.10
Information technology	5.90	2.60	2.40	1.00	4.8	2.50	3.00
Test evaluation	0.80	6.50	6.80	5.70	5.7	6.80	7.10
Auditing	0.00	4.50	3.00	0.40	0.6	4.10	1.00
Facilities engineering	2.80	4.20	4.40	4.30	2.9	1.90	1.00
Missing or other	30.70	11.30	13.70	18.50	11.9	7.40	1.00

entered a certain level of education, either via an education upgrade, as a new hire, or as a transfer in (Figures 2.2, 2.3, and 2.4). For example, Figure 2.2 shows that individuals who upgraded to a bachelor's degree in 1998 while working in the AW stayed for nine years after getting the upgrade, while new hires or transfers into the AW in 1998 with a bachelor's degree (and did not attain additional upgrades) separated after seven and six years, respectively. Across all three degrees and every year (with the exception of 2002 for a bachelor's), those who attained education upgrades had a longer time to separation than those who entered or transferred in.[10]

Figures 2.5, 2.6, and 2.7 plot yearly separation rates over the course of an individual's career. For individuals whose final education level was a bachelor's degree, new hires and transfers separate at a higher rate early on than education upgraders do, but after six to seven years, this is reversed. For individuals whose final education level is a master's or a PhD, the differences early on are very large and are likely to be important cumulatively, even though separation rates converge over time.

In addition, there is a difference between a worker who participates in a formal education program within DoD and one who is pursuing an education upgrade unsponsored. Participation in formal programs, which requires staying in the workforce to get support, may lead to artificially longer retention. Unfortunately, variables showing whether an individual received formal education assistance do not appear in the data until 2006. Table 2.7 compares characteristics of those who took advantage of three education support programs (Cooperative Education Program, Tuition Assistance or Reimbursement and Training Program, and Repayment of Student Loans Program) with the rest of the AW. Only a small percentage of the AW takes advantage of these programs. Participants generally have more experience and are more likely to be supervisors than the rest of the AW. Also, loan repayment may be fundamentally different, since the education is presumably obtained prior to joining the AW.

---

[10] Of course, this is not controlling for any of the other observables.

Figure 2.2
Years Until Separation, by Final Observed Education—Bachelor's

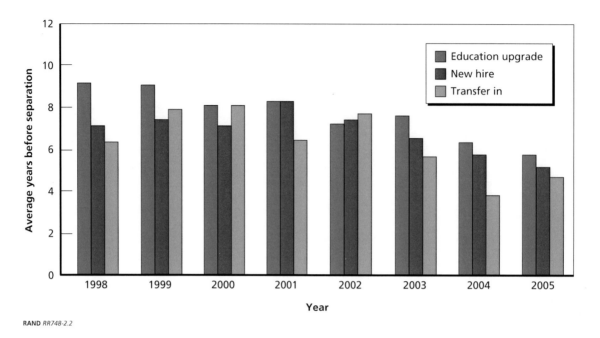

RAND *RR748-2.2*

Figure 2.3
Years Until Separation, by Final Observed Education—Master's

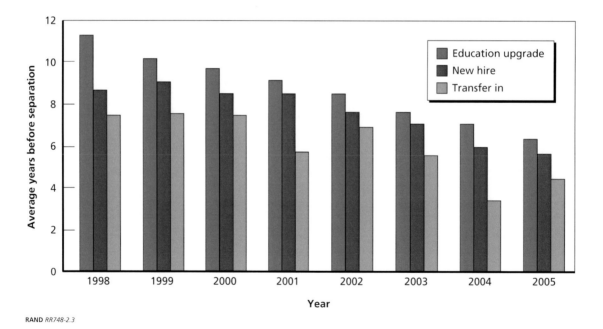

RAND *RR748-2.3*

Performance Rating

The finding that higher performance ratings increase the hazard of separation is robust across both specifications. We also conducted further analysis that allowed the effect of a higher performance rating on retention to vary depending on the seniority of the employee. Appendix B shows the results for two additional specifications, including interaction variables between

Figure 2.4
Years Until Separation, by Final Observed Education—PhD

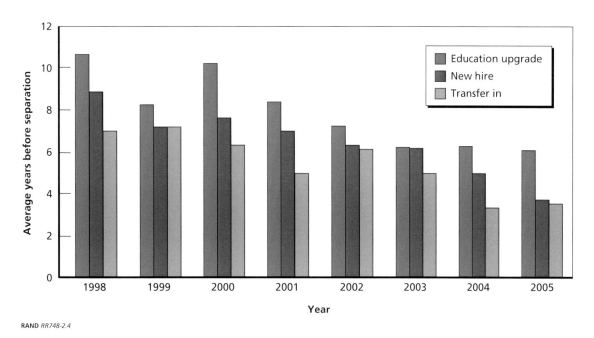

RAND *RR748-2.4*

Figure 2.5
Yearly Separation Rates, by Final Observed Education, FY 1998–2005 Cohorts—Bachelor's

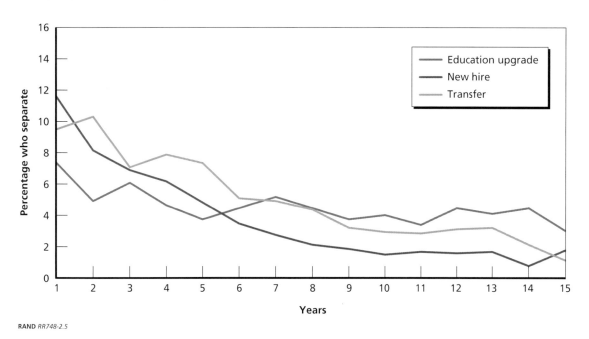

RAND *RR748-2.5*

grade and performance rating. For a low-grade employee, every one-unit increase in performance rating was associated with a small (10 percent) but significant increase in hazard of separation. However, for middle- and high-grade employees, an increase in performance rating led to a much larger (53 to 66 percent) increase in hazard of separation. This suggests that high-performance, high-grade employees are especially difficult to retain.

Figure 2.6
Yearly Separation Rates, by Final Observed Education, FY 1998–2005 Cohorts—Master's

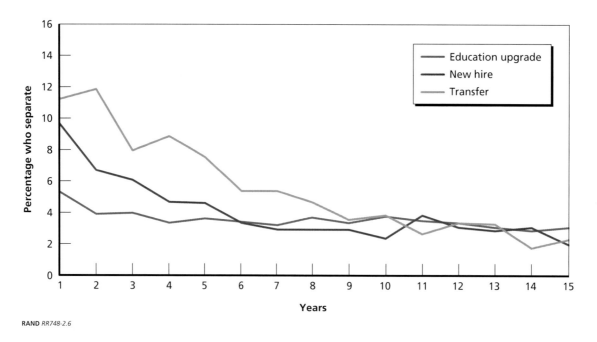

RAND *RR748-2.6*

Figure 2.7
Yearly Separation Rates, by Final Observed Education, FY 1998–2005 Cohorts—PhD

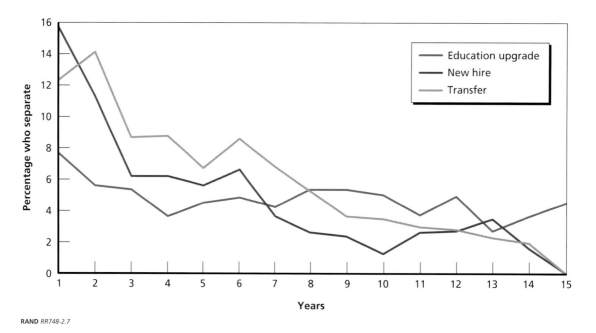

RAND *RR748-2.7*

During the course of this study, we experimented with additional specifications, replacing "average performance rating" with "initial performance rating" or with variables for the percentage of performance ratings at a certain level (e.g., percentage of "4" performance ratings, percentage of "5" performance ratings). Results were comparable. Furthermore, Appendix B includes Cox regressions of time to first promotion on performance rating, to verify the useful-

Table 2.7
Summary of Acquisition Workforce, by Type of Educational Assistance, FY 2013

Variable	Tuition Assistance	Coop	Repay Loan	Rest of AW
Average age	51.2	51.4	46.5	46.0
Average months of service	26.5	23.8	18.7	15.4
Average compensation ($)	93,824	92,195	75,003	80,601
% who are				
Veterans	6.1	25.0	11.1	29.8
Female	59.8	75.0	37.0	69.0
Supervisors	40.0	25.0	27.8	18.8
Education (%)				
Some college	15.0	16.7	13.0	8.2
Bachelor's	35.7	50.0	38.9	45.4
Master's	42.8	25.0	46.3	34.3
PhD	2.6	0.0	0.0	3.2
Grade (%)				
Entry	0.9	0.0	0.0	3.4
Middle	58.1	58.3	88.9	74.6
Senior	39.4	33.3	9.3	20.0
SES	0.5	0.0	1.9	0.4
Career field (%)				
Systems engineering	30.0	50.0	3.7	29.7
Program management	7.4	8.3	0.0	8.4
Contracting	36.4	16.7	81.5	18.6
Purchasing	1.0	0.0	0.0	0.9
Production quality	6.5	0.0	7.4	6.6
Business	2.0	0.0	0.0	4.7
Life-cycle logistics	5.7	25.0	1.9	11.9
Information tech	0.7	0.0	0.0	4.2
Test evaluation	8.2	0.0	0.0	5.0
Auditing	0.0	0.0	0.0	3.2
Facilities engineering	0.3	0.0	0.0	5.2
Missing or other	0.9	0.0	5.6	0.6
Individuals	767	12	54	134,460

ness of performance rating as an indicator of personnel quality. The third table of that appendix shows that higher performance ratings are associated with a higher likelihood of promotion.

Other Explanatory Variables

In general, the effects of the other explanatory variables in Table 2.5 are consistent across both specifications. Older and longer-tenured employees are more likely to separate because they are closer to retirement. Supervisors are much more likely to separate. Veterans are less likely to separate (15 percent greater chance of retention in a given year), which is in line with findings from Asch, 2001. The estimate for compensation indicates an extra $1,000 of income increases the probability of retention by 1.2 percent. Middle-grade employees have a lower hazard of separation than low-grade employees do. In specification (1), being a senior-grade employee does not significantly affect retention. However, when interactions between grade and performance rating are introduced, senior-grade employees have a lower hazard of separation than low-grade employees do. Employees who transferred into the AW have about a 40-percent lower separation rate than new hires do. One possible reason may be that AW managers and employees have more information to assess the degree of fit between the employee and the position in the case of an internal transfer into the AW. The race and gender of an employee were not related to retention after controlling for other factors.[11]

There is significant difference in separation hazard between branches of DoD. In particular, the Army has a significantly higher probability of separation than the Office of the Secretary of Defense (OSD) and the defense agencies do, while the Navy or Marines and Air Force were not significantly different from OSD. The difference might be tied to the culture of the military branch. Gates et al., 2013, notes that a majority of civilian AW hires with prior military experience in the Air Force, Army, Navy, or Marine Corps had their military service in the same organization. In contrast, OSD hires veterans from each of the services. Furthermore, the location of the position may affect retention because retention depends on the available supply of possible outside job prospects. This suggests a need for further understanding of losses specific to a given branch.

The hazard of separation varies significantly between career fields. In our analysis, systems engineering, which includes 27.2 percent of the AW, is the baseline career field to which others are compared. Most of the career fields have higher separation hazards than systems engineering does, especially program management and information technology. This may be due to stronger private-sector job prospects for individuals in these career fields. Life-cycle logistics is the only career field that has consistently lower hazards of separation than the baseline does.

Finally, keeping in mind that the year of entry into the AW may affect an individual's time to separation, we included dummy variables for FY 1999–FY 2005. Compared to 1998, all other years have lower hazards of separation, with the exception of 2004, which does not have a significant coefficient. Explanations for year-to-year variation in retention include variation in the types of jobs filled between years, differences in career or promotion prospects due to DoD funding, and the state of the U.S. economy (e.g., a strong economy might imply better private sector opportunities).

Interpretation of Results

In applying our findings to the current state of the AW, it is important to consider that the composition and characteristics of the AW have changed a bit over the past 15 years. Table 2.8 includes summary statistics for the earliest cohorts in our study (FYs 1998–2000) and the most recent cohorts entering the AW (2010–2012). Some differences between these two groups include more veterans, fewer women, fewer months of experience, and more education in the most recent cohorts. The percentage of workers for the Army has also increased substantially in the most recent cohort. One of our results found that higher-quality workers, especially senior, higher-quality workers in the Army, had drastically lower rates of retention. This may be of concern as the FY 2010–2012 cohorts progress through time and as the retention of the best employees in these cohorts becomes an issue.

---

[11] The explanatory variable differentiated only between white and nonwhite.

Table 2.8
Comparison of Earliest and Most Current Acquisition Workforce Cohorts

Variable	1998–2000	2010–2012
Average age	37.2	39.2
Average months of service	103.5	54.1
Average compensation ($000)	41.4	57.5
% who are		
Supervisors	6.3	5.1
Veterans	24.3	40.9
Female	42.3	32.6
Not white or Hispanic	25.6	27.2
Handicapped	7.2	7.4
Grade (%)		
Entry	45.3	29.2
Middle	51.3	62.4
Senior	3.4	7.8
Substantive transfer in (%)	46.6	30.4
Performance rating		
Average	3.8	4.2
% with a rating of		
1	0.0	0.1
2	0.1	0.2
3	39.9	15.8
4	33.3	19.1
5	22.9	34.1
Missing (%)	3.4	30.7[a]
Organization (%)		
Army	22.8	61.7
Navy or Marines	48.9	21.9
Air Force	14.5	1.3
OSD or other	13.8	15.1
Education (%)		
No college	14.0	17.2
Some college	11.5	7.9
Bachelor's degree	52.2	45.9
Master's degree	17.0	23.4
PhD	1.9	1.9
Career field (%)		
Systems engineering	25.0	27.0
Program management	5.0	7.2
Contracting	21.3	13.7
Purchasing	4.4	1.1
Production quality	3.5	5.9

Table 2.8—Continued

Variable	1998–2000	2010–2012
Business	9.1	7.3
Life-cycle logistics	11.7	15.2
Information technology	2.8	3.4
Auditing	5.1	5.5
Facilities engineering	6.3	4.6
Test evaluation	0.0	6.4
Missing or other	5.9	2.7
Individuals	4,234	10,209

[a] There is a greater incidence of missing performance ratings in the most recent years because individuals in their first year often do not receive performance ratings.

Conclusions

One main finding in this analysis is that a higher average performance rating is associated with an increased hazard of separation (decreased retention). Our results also imply that the effect of performance ratings on hazard of separation is much greater for employees who entered the AW at more-senior grades.

A second finding is with respect to how education, as a measure of personnel quality, affects the hazard of separation. Using initial education level as the explanatory variable, our results are similar to those of Asch, 2001, which also uses initial education: We found that retention is higher for people who enter the workforce with a bachelor's or master's degree than with those who enter the workforce with less than a bachelor's degree. We found no difference in retention between people who enter with a PhD and those who enter with less than a bachelor's degree. Using final education level as the explanatory variable, individuals last observed with a bachelor's, master's, or PhD are all more likely to be retained than those with less than a bachelor's degree. The intention to upgrade education appears to be an important indicator of retention, but it is not observed clearly. Based on exploratory analysis, our results also suggest, at least on average, that individuals who attain these degrees while in the workforce have lower separation rates. This finding implies that programs to encourage the AW to pursue higher education while in DoD service may have a strong positive effect on retaining quality workforce. Further analysis needs to be conducted to disentangle the effect of educational upgrades on retention, while addressing issues of confounding variables.[12] Additional research could address the question of whether it is better to hire individuals at a higher level of education directly or to bring individuals in at a lower level and eventually "groom" them with formal educational support programs.

---

[12] It is possible that those displaying the intention to pursue an educational upgrade are more likely to have underlying, unobservable characteristics, such as ambition and persistence, and that these characteristics also drive retention outcomes. In that case, better retention among upgraders may be overstating the effect of upgrading in that it may also be capturing the effects of ambition and persistence.

CHAPTER THREE

SES Profile Comparison

This chapter explores the careers and characteristics of the 498 SES employees who were part of the AW as of the end of FY 2012. We first present a descriptive summary of the SES to gain insight on how the AW leadership compares to the rest of the workforce. Then, we will present a logistic regression model that examines the factors that can help predict whether an individual will ever become SES.

Descriptive Overview of the SES Acquisition Workforce

We start by providing a brief overview of the acquisition workforce SES as of September 2012. Of the 498 SES, 417 were men and 433 were white. The median age was 53, with a range from 31 to 77. The SESs were employed in 38 different metropolitan regions in the United States, with 159 employed in the Washington, D.C., region.

The majority (262) of the AW SES personnel were employed by the Army. One hundred and thirty-eight worked for the Navy, 90 for OSD, and only eight for the Air Force. The median base SES salary was $163,000, with a range from $100,000 to $179,000.

Since 1992, the lowest number of SESs in the AW was 327 in 1994 (Figure 3.1). The number reached a peak of 575 in 2011 because there were 206 new hires that year, the greatest number of new SESs in a single year. Out of the 206 new SESs, 146 were in the general engineering occupation, and 186 were in the Army. Figure 3.2 compares the percentages of overall AW and SES by career field to highlight disparities between the two groups. SES workers are overrepresented in the AW, predominantly in the system engineering and program management career fields.

Factors That Predict Entering the SES Acquisition Workforce

This section examines the careers of the AW population in 1998 and the characteristics associated with a high probability of entering the SES. We chose 1998 as the point of comparison so that the period of observation would be adequately long, and more arbitrarily, for the sake of continuity with the previous sections. In 1998, there were 80,788 personnel in the AW,[1] excluding the SES. Of these, 590 were promoted to the SES by 2012, out of whom 319 cur-

---

[1] This number is based on historical data at DMDC and revised count methodologies after the later 1990s yielded higher numbers in the AW.

Figure 3.1
Acquisition Workforce SES Size, by Year

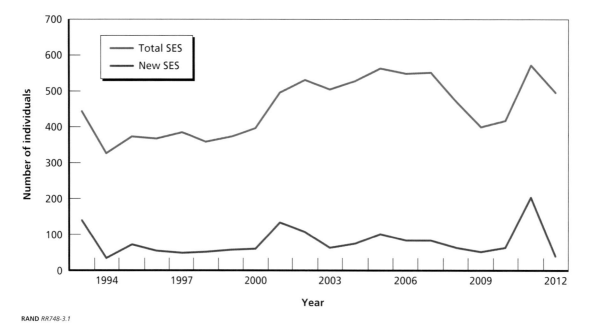

Figure 3.2
Percentage of AW and SES, by Career Field, September 2012

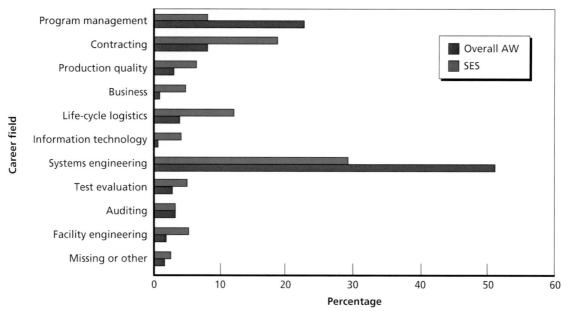

rently remain in their positions.[2] Table 3.1 shows that the vast majority of SES personnel as of the end of FY 2012 entered the workforce before 1998, with a small upswing of hires in 2002 and 2003.[3]

Table 3.2 compares all personnel in the AW in 1998 who did not eventually become SES with the personnel who did eventually become SES. The future SES leaders were unsurprisingly more senior (none of them were in an entry grade) and earned more money ($21,000 more than the rest of the AW). Fifty-seven percent were supervisors, compared to just 19 percent in the general workforce. The future SES was also highly educated; 10.9 percent had PhDs, and 56.3 percent had master's degrees. They were less likely to be veterans and more likely to be white males. Some career fields, such as program management and systems engineering, overproduced leaders, while contracting and life-cycle logistics had disproportionately few leaders. Forty-nine percent of the future SES was in the Army, despite the Army only making up 28 percent of the workforce.

We used a logistic regression model to further analyze the factors that increase the propensity for being a future leader (Table 3.3). Again, we chose the FY 1998 cohort and set the

Table 3.1
First Year in DoD for SES Members as of End of FY 2012

Year of Entry	Current SES
Prior to 1992	409
1993	2
1994	1
1995	5
1996	1
1997	2
1998	3
1999	3
2000	2
2001	3
2002	13
2003	14
2004	6
2005	4
2006	6
2007	6
2008	3
2009	9
2010	2
2011	3
2012	1

[2] At the end of 2012, 13 additional individuals out of the 590 remained in the AW but at lower grades.

[3] Current SES personnel who were hired after 1998 tend to be senior retired military. The model did not control for a veteran's rank, however, because the rank variable is often missing in DMDC's Work Experience File.

Table 3.2
Characteristics of Total Acquisition Workforce Population in FY 1998, by Non-SES and Future SES

Variable	Are Never SES	Future SES
Average age	45.97	41.86
Average months of service	224.29	213.04
Average compensation ($000)	54.24	75.53
% who are		
Supervisors	19.21	57.12
Veterans	20.17	8.64
Female	34.90	17.97
Not white or Hispanic	20.57	11.02
Handicapped	8.54	3.56
Education (%)		
Some college	15.70	1.36
Associate's	4.00	0.00
Bachelor's	37.50	30.85
Master's	26.17	56.27
PhD	3.89	10.85
Organization (%)		
Army	28.02	49.15
Navy or Marines	29.36	25.93
Air Force	20.70	15.25
OSD or other	21.92	9.66
Grade level (%)		
Entry	78.45	32.37
Middle	12.28	67.29
Senior	12.28	67.29
Career field (%)		
Systems engineering	27.85	44.06
Program management	5.68	18.98
Contracting	24.92	11.36
Production quality	7.23	3.90
Business	4.95	2.20
Life-cycle logistics	8.55	4.58
Information tech	2.14	1.02
Test evaluation	5.13	3.39
Auditing	4.20	3.22
Facilities engineering	4.20	3.22
Missing or other	5.15	4.07
Personnel	80,198	590

dependent variable equal to 1 for individuals who later entered the SES and 0 otherwise. Negative coefficient estimates indicate a negative relationship between the likelihood of entering the SES and the independent variable, while positive estimates indicate a positive relationship. For

Table 3.3
Logistic Regression of Future SES, FY 1998 Cohort

Variables	Estimate	Std. Error	Odds Ratio
Average age	−0.195***	−0.012	0.823***
Average months of service	−0.004***	0.001	0.996***
Average compensation ($000)	0.137***	0.007	1.147***
% who are			
Supervisors	0.463***	0.119	1.589***
Veterans	−0.022	0.161	0.978
Female	0.095	0.120	1.099
Not white or Hispanic	−0.077	0.139	0.926
Handicapped	−0.232	0.233	0.793
Education			
Some college	0.259	0.628	1.295
Bachelor's	0.589	0.524	1.802
Master's	1.066**	0.522	2.903**
PhD	0.933*	0.543	2.543*
Organization			
Army	0.955***	0.196	2.600***
Navy or Marines	0.001	0.202	1.001
Air Force	0.389*	0.216	1.476*
Grade			
Middle	2.006**	0.787	7.431**
Senior	2.614***	0.770	13.653***
Career fields			
Program management	0.291**	0.128	1.338**
Contracting	0.033	0.156	1.034
Production quality	0.916***	0.242	2.499***
Business	−0.688**	0.298	0.503**
Life-cycle logistics	0.285	0.218	1.330
Information tech	−0.213	0.432	0.808
Test evaluation	−0.901***	0.238	0.406***
Auditing	1.218***	0.316	3.380***
Missing or other	0.180	0.229	1.197
Constant	−8.441***	1.086	0.000***

NOTE: 77,582 observations.

*** $p < 0.01$, ** $p < 0.05$, * $p < 0.1$.

the most part, the regression model identified the same factors correlated with becoming part of the SES that the comparison of summary statistics did. The odds of a supervisor entering the SES were 59 percent higher than those of a nonsupervisor.[4] The odds of individuals in the

---

[4] Results from the logistic regression are interpreted in terms of odds as opposed to probabilities. High odds correspond to high probabilities, and low odds to low probabilities. An odd is the probability of success (entering the SES) over the probability of failure (not entering the SES).

Army reaching the SES are 160 percent higher than those of individuals in OSD. Compared to the baseline career field of systems engineering, individuals in production quality and program management are more likely to become part of the SES.

Gender was not a significant predictor of whether someone is promoted to the SES, despite the fact that 35 percent of the AW but only 18 percent of the SES are female. The reason for this discrepancy is that females tend to work in career fields that are underrepresented in the SES. For example, 21.3 percent of females in the AW work in the contracting career field, compared with only 9.6 percent of males. Personnel who work in contracting are less likely to ascend to senior levels, in part due to a high turnover rate; only 11.4 percent of the SES are in contracting. Over 50 percent of the SES consists of the systems engineering career field, but only 7.8 percent of AW females work in systems engineering (compared to 33.8 percent of males). As a result, the pool of women in career fields from which the SES recruits is smaller than that of men.

Race was also not a significant predictor in the regression model. However, only 11 percent of the SES are a race other than white, compared to 20.6 percent across the AW. Racial minorities are also underrepresented in some career fields, such as program management and systems engineering, that have a disproportionate number of SES. The effect is not as large as for women, however; 26 percent of minorities are in systems engineering, compared to 30 percent of whites.

It is important for the AW to devote resources to succession planning to ensure that it has future leaders in the pipeline to take over when the current SESs retire. While the SES can be externally hired, we observe that historically, nearly all of the SES in the AW are promoted from within. Most candidates for the SES are in senior grades, highly educated, and are clustered in several career fields, such as program management and systems engineering. Therefore, one way for DoD to meet future hiring demands of the SES is to maintain organically, a viable pool of individuals that meet these criteria.

CHAPTER FOUR

Effects of Acquisition Demonstration Pay Plan on Retention

This chapter applies the same survival analysis methodology to evaluate the effects of the AcqDemo on retention outcomes. As discussed in the introduction to this report, demonstration pay plans, including AcqDemo, were designed to provide additional flexibility in payment and promotion to workforce managers, with the hopes of improving retention outcomes (Werber et al., 2012; Campbell et al., 1993; Kettl et al., 1996; Johnston, 1988).

As with our analysis in Chapter Two, this analysis focuses on cohorts of new entrants to the AW in FYs 1998–2005. In this chapter, we account for the type of pay plan into which new employees entered.[1] We account for AcqDemo; other demonstration pay plans; and other non-GS, nondemonstration pay plans. Table 4.1 compares personnel covered by the four pay plan types.[2] AcqDemo personnel are, on average, older and more likely to be veterans than the other pay plans. The AcqDemo plan covered 4 percent of new AW civilian employees. Employees

Table 4.1
Comparison of AW, by Pay Plan Type, FY 1998–2005 Cohorts of New Hires

Variable	GS	AcqDemo	Other Demos	Other Pay Plans
Average age	37.8	41.9	36.3	36.3
Average months of service	97.8	79.1	62.3	105.4
Average compensation ($000)	46.1	61.1	57.4	59.2
% who are				
Supervisors	5.5	9.9	3.8	17.8
Veterans)	27.7	46.6	20.0	15.7
Female	37.2	30.8	25.5	17.9
Not white or Hispanic	24.4	25.6	19.2	20.0
Handicapped	8.0	7.8	6.1	8.8
Grade (%)				
Entry	36.5	28.7	41.5	22.0
Middle	60.2	57.3	52.9	63.6
Senior	3.2	14.0	5.4	14.4
Substantive transfer in	49.2	33.0	24.5	34.0

---

[1] Most of the AcqDemo personnel transitioned into and out of NSPS from 2007–2011 (Werber et al., 2012). The 858 personnel in our sample defined as AcqDemo on entry include both the ones who transitioned and the ones who stayed in AcqDemo the entire time.

[2] The comparison is by the pay plan under which they were initially hired.

Table 4.1—Continued

Variable	GS	AcqDemo	Other Demos	Other Pay Plans
Performance rating				
Average	3.9	4.0	4.1	3.4
% with a rating of				
1	0.0	0.0	0.2	0.1
2	0.1	0.1	0.5	0.1
3	25.5	17.4	26.8	63.6
4	26.9	63.6	31.7	17.1
5	41.6	18.1	39.8	9.3
Organization (%)				
Army	53.6	60.3	52.8	7.4
Navy or Marines	32.9	16.3	37.2	91.3
Air Force	6.7	22.7	9.9	0.8
OSD or other	6.9	0.7	0.0	0.5
Education (%)				
No college	16.8	17.8	8.2	3.5
Some college	11.4	7.3	5.0	3.4
Bachelor's	50.7	37.6	53.1	66.6
Master's	16.3	28.6	24.1	21.3
PhD	1.4	3.3	8.0	4.6
Career field (%)				
	22.7	11.4	82.0	64.7
Systems engineering	7.1	23.3	3.3	5.1
Program management	17.3	9.9	1.7	3.2
Contracting Purchasing	2.2	0.7	0.1	0.2
Production quality)	4.3	0.1	0.3	4.2
Business	8.7	13.1	0.9	5.8
Life-cycle logistics	15.3	9.6	2.4	3.2
Information technology	4.1	8.0	1.4	5.3
Auditing	5.1	21.1	6.2	4.8
Facilities engineering	3.9	0.0	0.0	0.0
Test evaluation Missing or	4.7	0.1	0.3	0.9
other	4.7	2.7	1.6	2.7
Individuals	18,342	858	1,942	1,922

on the AcqDemo plan tend to be older, more likely to be veterans, and more highly paid than the AW civilian employees on other pay plans. The distribution of performance ratings differs between those on AcqDemo and those on the other pay plans.

We devised two specifications (Table 4.2) to examine the impact of AcqDemo, along with other explanatory variables, on the dependent variable—time to separation. First, we treated AcqDemo and the other demonstration plans separately when comparing retention outcomes to the GS plan. Specification (1) of the Cox regression includes dummy variables for whether someone entered the AW on the AcqDemo plan, another demonstration pay plan, or

Table 4.2
Time to Separation Cox Regressions, FY 1998–2005 Cohorts

| Variables | Time to Separation | | | | | |
| | Specification (1) | | | Specification (2) | | |
	Estimate	Std. Error	Hazard Ratio[a]	Estimate	Std. Error	Hazard Ratio[a]
Performance rating	0.304***	0.030	1.355***	0.304***	0.030	1.355***
AcqDemo	−0.268***	0.087	0.765***			
Other demo	−0.168***	0.061	0.845***			
Other pay plan	0.232***	0.065	1.261***			
All demo pay plans				−0.227***	0.051	0.797***
Average age	0.020***	0.002	1.020***	0.020***	0.002	1.020***
Average months of service	0.004***	0.000	1.004***	0.004***	0.000	1.004***
Compensation ($000)	−0.009***	0.001	0.991***	−0.009***	0.001	0.991***
Those who are						
Supervisors	0.245***	0.054	1.277***	0.267***	0.053	1.307***
Veterans	−0.165***	0.036	0.848***	−0.170***	0.036	0.844***
Female	−0.009	0.035	0.991	−0.008	0.035	0.992
Not white or Hispanic	−0.019	0.033	0.981	−0.020	0.033	0.980
Handicapped	0.073	0.049	1.076	0.077	0.049	1.080
Organization						
Army	0.234**	0.091	1.263**	0.219**	0.091	1.245**
Navy or Marines	−0.079	0.090	0.924	−0.059	0.090	0.942
Air Force	−0.117	0.107	0.889	−0.129	0.106	0.879
Grade						
Middle	−0.136***	0.044	0.873***	−0.139***	0.044	0.870***
Senior	−0.043	0.087	0.958	−0.030	0.087	0.970
Substantive transfer in	−0.554***	0.032	0.575***	−0.566***	0.032	0.568***
Education						
Bachelor's	−0.164***	0.036	0.849***	−0.156***	0.036	0.855***
Master's	−0.155***	0.045	0.857***	−0.148***	0.045	0.862***
PhD	0.142	0.089	1.152	0.153*	0.089	1.165*
Career field						
Program management	0.635***	0.057	1.888***	0.591***	0.056	1.806***
Contracting	0.112*	0.058	1.119*	0.071	0.056	1.073
Purchasing	0.383***	0.099	1.467***	0.344***	0.099	1.411***
Production quality	0.252***	0.076	1.286***	0.220***	0.075	1.246***
Business	0.443***	0.059	1.557***	0.399***	0.057	1.490***
Life-cycle logistics	−0.272***	0.061	0.762***	−0.313***	0.060	0.731***
Information tech	0.793***	0.065	2.210***	0.751***	0.064	2.118***
Test evaluation	−0.038	0.085	0.963	−0.079	0.084	0.924
Auditing	0.826***	0.129	2.285***	0.785***	0.128	2.192***
Facilities engineering	0.859***	0.075	2.362***	0.795***	0.073	2.215***
Missing or left out CF	0.635***	0.057	1.888***	0.591***	0.056	1.806***

Table 4.1—Continued

| Variables | Time to Separation | | | | | |
| | Specification (1) | | | Specification (2) | | |
	Estimate	Std. Error	Hazard Ratio[a]	Estimate	Std. Error	Hazard Ratio[a]
Year dummies						
1999	−0.223***	0.075	0.800***	−0.218***	0.075	0.804***
2000	−0.528***	0.081	0.590***	−0.530***	0.081	0.589***
2001	−0.031	0.060	0.970	−0.018	0.060	0.982
2002	−0.441***	0.065	0.644***	−0.424***	0.065	0.655***
2003	−0.350***	0.075	0.705***	−0.340***	0.075	0.712***
2004	−0.257***	0.068	0.773***	−0.247***	0.068	0.781***
2005	−0.481***	0.069	0.618***	−0.474***	0.069	0.623***
Censored observations	6,139	6,139	6,139	6,139	6,139	6,139
Observations	21,714	21,714	21,714	21,714	21,714	21,714

[a] Hazard Ratio = exp(Estimate).

another non-GS, nondemonstration pay plan. According to the hazard ratio of 0.765, retention increases by 24 percent (1 − 0.765) due to entering the workforce on the AcqDemo pay plan rather than the GS plan. Entering on another demonstration plan increases retention by 16 percent, and entering on the other pay plans decreases retention by 26 percent over the GS plan. In specification (2), we bundled AcqDemo in with the other demonstration pay plans, comparing all demonstration plans with the GS plan. According to the risk ratio of 0.797 for the All Demo Plans dummy variable, retention increases by 20 percent over the GS plan, while all other results remain essentially the same across both specifications.

Therefore, it is noteworthy that, in both specifications, people who were in AcqDemo and, in fact, any demonstration pay plan were retained longer than those in the GS plan.[3] Nevertheless, it is possible that unobserved confounding factors may be influencing both entry into AcqDemo and retention outcomes. Failure to account for such factors could result in mis-estimation of the perceived relationship between AcqDemo and retention. One possible factor is that that employees in AcqDemo had greater compensation growth than otherwise-similar non-AcqDemo employees (Werber et al., 2012, Appendix B). Preliminary findings suggest that accelerated compensation, particularly early in a career, will affect retention (Feintzeig, 2014).

---

[3]   In Appendix C, we show that this result holds using percentage of time in one's career spent in AcqDemo as the explanatory variable.

CHAPTER FIVE

Conclusion

An analysis of workforce retention can shed light on retention rates and how they change over time. When analyzing workforce retention, DoD managers seek to understand not only whether workers are staying but also whether the right workers are staying, who is being promoted to leadership positions, and what organizational factors might influence retention.

In this report, we have reviewed some plausible and readily available measures of workforce quality and analyzed the relationship between these measures and workforce retention. We considered two quality metrics: average performance ratings and educational attainment. We also analyzed promotion into SES positions and considered whether individuals who are hired into the AcqDemo pay plan or other demonstration project pay plans are retained at different rates.

We found that different quality metrics have different relationships with retention. Looking first at average performance ratings, employees with a higher average performance rating were more likely to separate from DoD. Our results also imply that the effect of performance ratings on retention is much greater for employees who entered the AW at more-senior grades: Higher-rated individuals, especially the more senior ones, are more likely to leave DoD, presumably to pursue more-favorable career opportunities.

Educational attainment shows the reverse relationship. The probability of retention was greater for those with bachelor's or master's degrees at the time of entry than for those without a college degree at entry. Individuals last observed with a bachelor's, master's, or PhD are more likely to be retained than those with less than a bachelor's degree, and the difference is larger than in the analysis that used education levels at entry. The intention to upgrade education appears to be an important indicator of retention, but it is not observed clearly. Based on exploratory analysis, our results also suggest, at least on average, individuals who attain these degrees while in the workforce have lower separation rates.

Our analysis of promotion to the SES reveals that organizational characteristics are associated with promotion but that the demographic characteristics of the employee are not. Individuals with an Army background account for almost one-half the SES ranks, even though they make up just 28 percent of the AW. In contrast, OSD personnel make up 22 percent of the AW but account for less than 10 percent of the SES ranks. Compared to the baseline career field of systems engineering, individuals in production quality, auditing, and program management are more likely to become part of the SES. Individuals in the business career field are less likely to become part of the SES. Gender and race were not significant predictors of whether someone is promoted to the SES, despite the fact that both groups are underrepresented in the AW SES relative to their prevalence in the overall AW. The reason for this discrepancy is that women and minorities tend to work in career fields that are underrepresented in the SES.

Regarding the third topic, our analysis provides evidence that people who entered the AW and were covered by the AcqDemo or, in fact, any demonstration pay plan were retained longer than those in the GS plan.[1] Retention was 24 percent higher for the AcqDemo pay plan than for the GS plan. Relative to the GS plan, entering on another demonstration plan increased retention by 12 percent, and entering on the other pay plans decreased retention by 33 percent.

This analysis provides a first look at some factors that are related to retention and promotion for civilians in the AW. Our analysis raises some concerns for DoD managers, in that we found higher rates of separation among workers with higher performance ratings. However, we also found that higher educational attainment is associated with lower rates of separation. Further investigation of the effect of educational upgrades while in the workforce on retention could be fruitful. The intention to pursue an educational upgrade is likely to be related to underlying unobserved characteristics of worker quality, such as ambition and persistence, and these characteristics are important in driving retention outcomes. Unfortunately, our data do not observe intention but rather report actual degree attainment during a limited period of observation. In addition, there is a difference between a worker who participates in a formal education program within DoD and one who is pursuing an education upgrade unsponsored. More research would address the question of whether it is better to hire for a certain level of education directly or to bring individuals in at a lower level and eventually "groom" them with formal educational support programs. Further analysis, including more-comprehensive metrics, would improve understanding of the dynamic between workforce quality and retention.

---

[1]   In Appendix C, we show that this result holds using percentage of time in one's career spent in AcqDemo as the explanatory variable.

Kaplan-Meier Survival Functions

The outcome of interest in this analysis is the length of stay until separation from the AW. In determining what explanatory variables to include in the Cox regression model of months until separation, we conducted univariate analysis of the effect of individual predictors on survival. For categorical predictors, we compared survival in different groups by plotting Kaplan-Meier curves and using log-rank significance tests. Each Kaplan-Meier curve shows a group-specific survival function over time. The survival function, which is related to the hazard function, is the probability of surviving to a certain time or beyond. Plotting the survival function over time for each group allows us to verify that the relative survival rates between groups do not change dramatically over time (i.e., the groups are proportional). However, this is only informative if there are sufficient observations for each value of a categorical predictor: As the size of each stratum decreases, the estimated survival functions become less precise. A common threshold for including a predictor is whether the log-rank test of significance can reject the null hypothesis with a probability of 0.2 or less; otherwise, it is highly unlikely that the predictor will contribute to a model including multiple other predictors. For continuous predictors, it is uninformative to generate a curve for each of the many possible levels. Instead, we used a univarate Cox proportional hazard regression.

The figures below show the results of univariate analysis on measures of personnel quality for the FY 1998 through FY 2005 cohorts. These Kaplan-Meier curves indicate the probability of survival (not leaving the workforce) stratified by education level on entry, final education level, and average performance rating over a career. The probability of survival (y-axis) decreases over time spent in the workforce (x-axis); across categories, these probabilities are roughly proportional and independent of time (i.e., the resulting curves are roughly parallel). In Figures A.1 and A.2, which show education on entry and final observed education, respectively, it is difficult to draw conclusions from the curve for PhD, due to the relatively small number of observations (only about 2 percent of the AW has a PhD). In Figure A.3, we did not plot curves for the lowest performance rating because there were not enough observations with an average performance rating of 1 or 2. Figures A.4 through A.7 provide additional univariate analysis of other predictors (e.g., entry grade, gender, supervisor status, veteran status). Again, the intent of univariate analysis is to take a preliminary exploration of the data and decide what explanatory variables to include in the multivariate Cox regression. By definition, it takes only one variable into account at a time, so to truly understand how multiple covariates together contribute to the outcome variable, we needed to run a Cox regression.

Figure A.1
Cumulative Probability of Retention, by Education on Entry

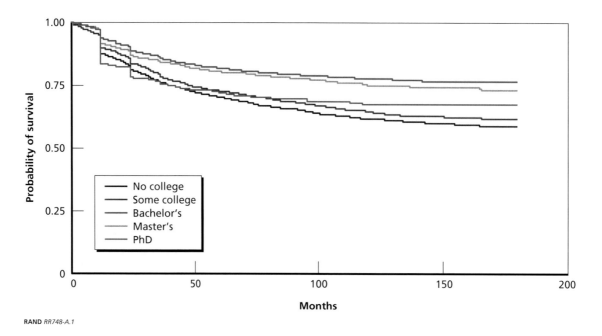

RAND *RR748-A.1*

Figure A.2
Cumulative Probability of Retention, by Final Education

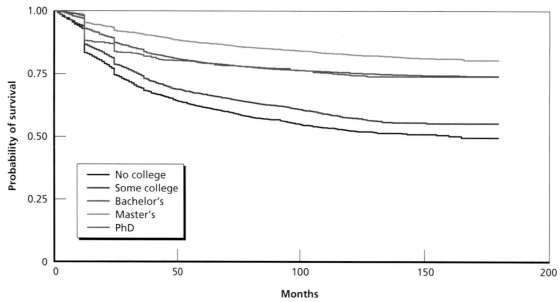

RAND *RR748-A.2*

Figure A.3
Cumulative Probability of Retention, by Average Performance Rating

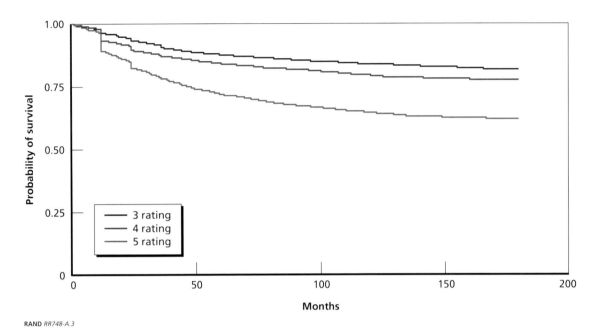

Figure A.4
Cumulative Probability of Retention, by Entry Grade

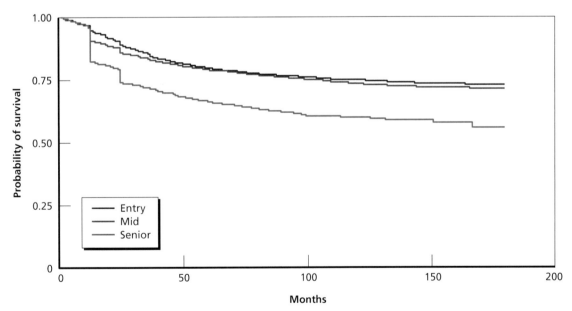

Figure A.5
Cumulative Probability of Retention, by Veteran Status

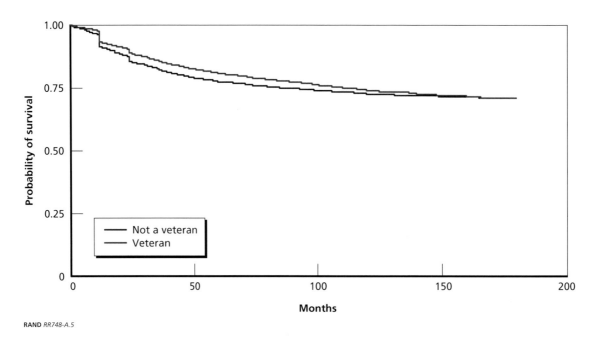

Figure A.6
Cumulative Probability of Retention, by Gender

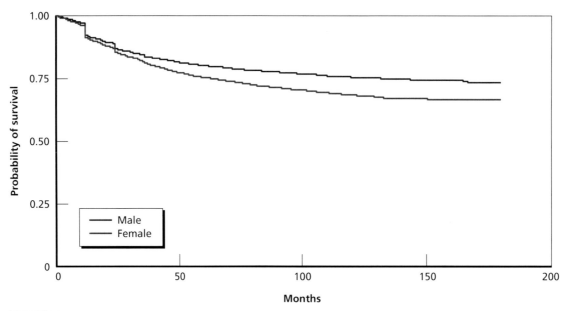

Figure A.7
Cumulative Probability of Retention, by Supervisor Status

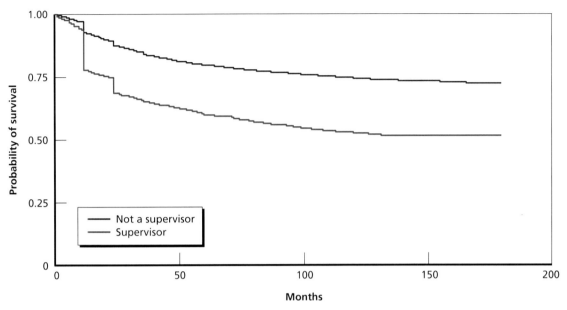

Robustness Checks

Specification (3) of Table B.1 interacts average performance rating with the entry grade of the employee, with the assumption that the effect of performance rating on retention will vary depending on the three possible grades (low, middle, senior). The regression also includes education on entry. The interpretation of coefficient estimates associated with the performance rating is now more complex than simply taking the exponential directly:

- If the employee has a low grade, hazard ratio = $\exp(0.095)$ = 1.10 (10-percent increase in hazard of separation for every one unit increase in performance rating).
- If the employee has a middle grade, hazard ratio = $\exp(0.095 + 0.331)$ = 1.53 (53-percent increase in hazard of separation for every one unit increase in performance rating).
- If the employee has a high grade, hazard ratio = $\exp(0.095 + 0.409)$ = 1.66 (66-percent increase in hazard of separation for every one unit increase in performance rating).

As noted earlier, the average performance rating varied across services. Specification (4) is similar to (3), except final education is used. Results for both are presented in Table B.2. Table B.3 shows that time to first promotion is significantly related to performance rating.

Table B.1
Four Specifications of Cox Regressions

	Average Performance Rating	
	Education	
	Upon Entry	Final Observed
No interaction	(1)	(2)
Interaction with grade	(3)	(4)

Table B.2
Cox Regression Model of Months Until Separation Including Interaction Terms, FY 1998–2005 Cohorts—Specifications (3) and (4)

| | Time to Separation | | | | | |
| | Specification (3) | | | Specification (4) | | |
Variables	Estimate	Std. Error	Hazard Ratio[a]	Estimate	Std. Error	Hazard Ratio[a]
Perf*midgrade	0.331***	0.042	1.393***	0.327***	0.042	1.386***
Perf*senior grade	0.409***	0.090	1.505***	0.403***	0.090	1.496***
Average age	0.020***	0.002	1.020***	0.019***	0.002	1.019***
Average months of service	0.004***	0.000	1.004***	0.004***	0.000	1.004***
Compensation ($000)	−0.010***	0.001	0.990***	−0.007***	0.001	0.993***
Those who are						
Supervisors	0.276***	0.053	1.318***	0.295***	0.053	1.344***
Veterans	−0.172***	0.036	0.842***	−0.162***	0.036	0.850***
Female	−0.019	0.035	0.981	−0.029	0.035	0.972
Not white or Hispanic	−0.024	0.033	0.976	−0.014	0.033	0.986
Handicapped	0.081*	0.049	1.084*	0.084*	0.049	1.087*
Organization						
Army	0.216**	0.091	1.242**	0.240***	0.091	1.272***
Navy or Marines	−0.026	0.090	0.975	−0.043	0.090	0.958
Air Force	−0.146	0.106	0.864	−0.046	0.107	0.955
Grade						
Middle	−1.499***	0.182	0.223***	−1.509***	0.182	0.221***
Senior	−1.732***	0.410	0.177***	−1.696***	0.408	0.183***
Substantive transfer in	−0.553***	0.032	0.575***	−0.557***	0.032	0.573***
Initial education						
Bachelor's	−0.151***	0.036	0.860***			
Master's	−0.143***	0.045	0.867***			
PhD	0.142	0.089	1.152			
Final education						
Bachelor's				−0.304***	0.038	0.738***
Master's				−0.652***	0.044	0.521***
PhD				−0.320***	0.087	0.726***
Career field						
Program management	0.579***	0.056	1.785***	0.528***	0.056	1.695***
Contracting	0.089	0.056	1.093	0.127**	0.056	1.136**
Purchasing	0.334***	0.099	1.397***	0.254***	0.098	1.290***
Production quality	0.233***	0.075	1.263***	0.142*	0.076	1.153*
Business	0.385***	0.058	1.469***	0.330***	0.058	1.391***
Life-cycle logistics	−0.331***	0.060	0.718***	−0.406***	0.060	0.666***
Information tech	0.741***	0.064	2.098***	0.626***	0.064	1.871***
Test evaluation	−0.095	0.084	0.909	−0.099	0.084	0.906
Auditing	0.796***	0.128	2.217***	0.828***	0.128	2.289***

Table B.2—Continued

	Time to Separation					
	Specification (3)			Specification (4)		
Variables	Estimate	Std. Error	Hazard Ratio[a]	Estimate	Std. Error	Hazard Ratio[a]
Facilities engineering	0.811***	0.073	2.251***	0.801***	0.073	2.228***
Missing or left out CF	0.727***	0.063	2.068***	0.674***	0.064	1.961***
Year dummies						
1999	−0.225***	0.075	0.799***	−0.229***	0.075	0.795***
2000	−0.521***	0.082	0.594***	−0.536***	0.082	0.585***
2001	−0.015	0.060	0.985	−0.045	0.060	0.956
2002	−0.396***	0.065	0.673***	−0.428***	0.065	0.652***
2003	−0.313***	0.075	0.731***	−0.329***	0.076	0.720***
2004	−0.221***	0.068	0.802***	−0.274***	0.068	0.760***
2005	−0.442***	0.069	0.643***	−0.482***	0.069	0.618***
Censored observations	6,139	6,139	6,139	6,139	6,139	6,139
Observations	21,714	21,714	21,714	21,714	21,714	21,714

[a] Hazard Ratio = exp(Estimate).

Table B.3
Time to First Promotion Correlated with Performance Rating: Cox Regression Time to Promotion, FY 1998–2000 Cohorts

Variables	(1) Time to 1st Promotion	(2) Time to 1st Promotion
Average age	1.042***	1.043***
	(0.004)	(0.004)
Months federal service at entry	1.002***	1.002***
	(0.000)	(0.000)
Compensation ($000)	1.014***	1.014***
	(0.005)	(0.005)
1 if supervisor	1.056	1.053
	(0.087)	(0.087)
1 if vet	1.082	1.083
	(0.069)	(0.069)
Female	1.065	1.071
	(0.070)	(0.071)
Not white or Hispanic	1.065	1.060
	(0.063)	(0.063)
Handicapped	1.166*	1.158*
	(0.092)	(0.091)

Table B.3—Continued

Variables	(1) Time to 1st Promotion	(2) Time to 1st Promotion
Performance ratings		
Average		1.223***
		(0.064)
Percentage that are		
4	1.529***	
	(0.165)	
5	1.460***	
	(0.155)	
Education		
Some college	0.935	0.933
	(0.068)	(0.068)
Associate's	0.922	0.924
	(0.103)	(0.103)
Bachelor's	0.755***	0.759***
	(0.058)	(0.058)
Master's	0.666***	0.674***
	(0.061)	(0.062)
PhD	0.677**	0.687**
	(0.124)	(0.126)
Starting grade		
6	0.816*	0.803*
	(0.096)	(0.094)
7 or 8	0.408***	0.399***
	(0.054)	(0.053)
9 or 10	0.487***	0.475***
	(0.070)	(0.067)
11	0.419***	0.405***
	(0.055)	(0.052)
12	0.355***	0.345***
	(0.056)	(0.055)
13	0.354***	0.342***
	(0.072)	(0.069)
14	0.338***	0.327***
	(0.092)	(0.089)
15	0.277***	0.267***
	(0.099)	(0.096)
Organization		
Army	0.690***	0.679***
	(0.066)	(0.065)
Navy	0.968	0.939
	(0.083)	(0.079)

Table B.3—Continued

Variables	(1) Time to 1st Promotion	(2) Time to 1st Promotion
Air Force	1.050	1.058
	(0.113)	(0.114)
Year dummies		
1999	1.042	1.035
	(0.072)	(0.072)
2000	1.047	1.045
	(0.073)	(0.073)
Observations	5,560	5,560

*** $p < 0.01$, ** $p < 0.05$, * $p < 0.1$.

AcqDemo Robustness

Given the challenges in analysis of AcqDemo, we devised three specifications (Table C.1) to determine the impact of AcqDemo, along with other explanatory variables, on the dependent variable, time to separation. Specification (1) of the Cox regression includes a dummy for whether someone entered the AW on the AcqDemo plan; 3.65 percent of AW civilian employees fell into this category. Employees on the AcqDemo plan tended to be older, more experienced, more highly educated, and more highly paid than typical AW civilian employees. According to the hazard ratio of 0.712, entering the workforce on the AcqDemo pay plan increased retention by 29 percent (1 – 0.712).

In specification (2), we used the same dummy as (1) but also controlled for performance ratings and the associated interaction variables. According to the risk ratio of 0.827, retention increased by 17 percent.

For specification (3), we included a variable defined as the percentage of career time spent on the AcqDemo plan. The mean of this variable was 4.89 percent across the AW data set, and the average time on the AcqDemo plan for those enrolled was 2.6 years. Because this variable has a range of 0 to 1, the interpretation for the hazard ratio of 0.317 is that someone who spent all their time in AcqDemo (percent years = 1) had a 68 percent greater chance of retention than someone who was never in AcqDemo (percent years = 0).

Table C.1
Time to Separation Cox Regressions, FY 1998–2005 Cohorts

Variables	Specification (1)		Specification (2)		Specification (3)	
	Estimate	Hazard Ratio	Estimate	Hazard Ratio	Estimate	Hazard Ratio
Average performance rating			0.086**	1.090**	0.039	1.039
			(0.043)	(0.046)	(0.043)	(0.044)
Average perf rating*midgrade			0.330***	1.391***	0.322***	1.380***
			(0.044)	(0.062)	(0.044)	(0.061)
Average perf rating*senior grade			0.418***	1.519***	0.372***	1.451***
			(0.094)	(0.142)	(0.092)	(0.134)
AcqDemo at entry	−0.340***	0.712***	−0.190**	0.827**		
	(0.085)	(0.060)	(0.087)	(0.072)		
Percentage of years in AcqDemo					−1.148***	0.317***
					(0.121)	(0.038)
Average age	0.020***	1.021***	0.021***	1.021***	0.020***	1.021***
	(0.002)	(0.002)	(0.002)	(0.002)	(0.002)	(0.002)

Table C.1—Continued

Variables	Specification (1)		Specification (2)		Specification (3)	
	Estimate	Hazard Ratio	Estimate	Hazard Ratio	Estimate	Hazard Ratio
Average months of service	0.004***	1.004***	0.004***	1.004***	0.004***	1.004***
	(0.000)	(0.000)	(0.000)	(0.000)	(0.000)	(0.000)
Compensation ($000)	−0.009***	0.991***	−0.011***	0.990***	−0.009***	0.991***
	(0.001)	(0.001)	(0.001)	(0.001)	(0.001)	(0.001)
Those who are						
Supervisors	0.303***	1.355***	0.313***	1.368***	0.303***	1.355***
	(0.056)	(0.075)	(0.056)	(0.076)	(0.056)	(0.075)
Veterans	−0.150***	0.860***	−0.161***	0.851***	−0.150***	0.860***
	(0.037)	(0.032)	(0.037)	(0.032)	(0.037)	(0.032)
Female	0.040	1.041	0.025	1.026	0.040	1.041
	(0.035)	(0.037)	(0.035)	(0.036)	(0.035)	(0.037)
Not white or Hispanic	−0.027	0.973	−0.028	0.972	−0.027	0.973
	(0.034)	(0.033)	(0.034)	(0.033)	(0.034)	(0.033)
Handicapped	0.067	1.070	0.066	1.068	0.067	1.070
	(0.051)	(0.055)	(0.051)	(0.055)	(0.051)	(0.055)
Education						
Bachelor's	−0.187***	0.830***	−0.176***	0.839***	−0.187***	0.830***
	(0.038)	(0.031)	(0.038)	(0.032)	(0.038)	(0.031)
Master's	−0.141***	0.868***	−0.142***	0.868***	−0.141***	0.868***
	(0.047)	(0.041)	(0.047)	(0.041)	(0.047)	(0.041)
PhD	0.128	1.137	0.138	1.147	0.128	1.137
	(0.091)	(0.104)	(0.091)	(0.105)	(0.091)	(0.104)
Organization						
Army	−0.246***	0.782***	−0.316***	0.729***	−0.246***	0.782***
	(0.085)	(0.067)	(0.085)	(0.062)	(0.085)	(0.067)
Navy or Marines	−0.513***	0.599***	−0.472***	0.624***	−0.513***	0.599***
	(0.086)	(0.051)	(0.086)	(0.053)	(0.086)	(0.051)
Air Force	−0.463***	0.629***	−0.480***	0.619***	−0.463***	0.629***
	(0.100)	(0.063)	(0.100)	(0.062)	(0.100)	(0.063)
Grade						
Middle	−1.523***	0.218***	−1.546***	0.213***	−1.523***	0.218***
	(0.192)	(0.042)	(0.193)	(0.041)	(0.192)	(0.042)
Senior	−1.635***	0.195***	−1.823***	0.161***	−1.635***	0.195***
	(0.419)	(0.082)	(0.425)	(0.069)	(0.419)	(0.082)
Substantive transfer in	−0.534***	0.586***	−0.491***	0.612***	−0.519***	0.595***
	(0.031)	(0.018)	(0.032)	(0.019)	(0.032)	(0.019)
Career field						
Program management	0.427***	1.533***	0.381***	1.463***	0.441***	1.554***
	(0.059)	(0.091)	(0.062)	(0.091)	(0.062)	(0.097)
Contracting	−0.150***	0.861***	−0.270***	0.763***	−0.257***	0.773***

Table C.1—Continued

Variables	Specification (1)		Specification (2)		Specification (3)	
	Estimate	Hazard Ratio	Estimate	Hazard Ratio	Estimate	Hazard Ratio
	(0.055)	(0.047)	(0.062)	(0.047)	(0.062)	(0.048)
Purchasing	0.031	1.031	0.006	1.006	0.020	1.020
	(0.103)	(0.106)	(0.110)	(0.110)	(0.110)	(0.112)
Production quality	−0.179**	0.836**	−0.174*	0.840*	−0.175*	0.839*
	(0.090)	(0.075)	(0.093)	(0.078)	(0.093)	(0.078)
Business	0.141**	1.152**	0.073	1.075	0.113*	1.120*
	(0.063)	(0.073)	(0.068)	(0.073)	(0.068)	(0.076)
Life-cycle logistics	−0.506***	0.603***	−0.631***	0.532***	−0.612***	0.542***
	(0.062)	(0.038)	(0.068)	(0.036)	(0.068)	(0.037)
Information tech	0.594***	1.811***	0.549***	1.731***	0.568***	1.766***
	(0.072)	(0.130)	(0.077)	(0.133)	(0.077)	(0.135)
Test evaluation	−0.280***	0.756***	−0.311***	0.733***	−0.269***	0.764***
	(0.088)	(0.066)	(0.099)	(0.072)	(0.098)	(0.075)
Missing or left out CF	0.577***	1.781***	0.529***	1.697***	0.528***	1.696***
	(0.048)	(0.086)	(0.052)	(0.088)	(0.052)	(0.088)
Year dummies						
1999						
2000	−0.431***	0.650***	−0.473***	0.623***	−0.472***	0.623***
	(0.074)	(0.048)	(0.081)	(0.050)	(0.081)	(0.050)
2001	−0.036	0.965	−0.034	0.967	−0.031	0.970
	(0.060)	(0.058)	(0.063)	(0.061)	(0.063)	(0.061)
2002	−0.218***	0.804***	−0.211***	0.810***	−0.215***	0.806***
	(0.061)	(0.049)	(0.066)	(0.053)	(0.066)	(0.053)
2003	−0.132*	0.876*	−0.080	0.923	−0.088	0.915
	(0.072)	(0.063)	(0.076)	(0.070)	(0.076)	(0.070)
2004	0.044	1.045	0.096	1.101	0.076	1.079
	(0.063)	(0.066)	(0.067)	(0.074)	(0.068)	(0.073)
2005	−0.273***	0.761***	−0.254***	0.775***	−0.268***	0.765***
	(0.065)	(0.050)	(0.071)	(0.055)	(0.071)	(0.054)
Individuals	20,554	20,554	20,554	20,554	20,554	20,554

*** $p < 0.01$, ** $p < 0.05$, * $p < 0.1$.

References

5 U.S. Code 4703, Demonstration Projects.

Asch, Beth J., *The Pay, Promotion, and Retention of High-Quality Civil Service Workers in the Department of Defense*, Santa Monica, Calif.: RAND Corporation, MR-1193-OSD, 2001. As of September 11, 2014: http://www.rand.org/pubs/monograph_reports/MR1193.html

Campbell, Alan K., Stephen J. Lukasik, and Michael G. McGeary, eds., *Improving the Recruitment, Retention, and Utilization of Federal Scientists and Engineers*, Washington, D.C.: National Academy Press, 1993.

Feintzeig, Rachel, "When One Pay Raise a Year Isn't Enough," *Wall Street Journal*, July 15, 2014.

Gates, Susan M., Edward G. Keating, Adria Jewell, Lindsay Daugherty, Bryan Tysinger, and Ralph Masi, *The Defense Acquisition Workforce: An Analysis of Personnel Trends Relevant to Policy, 1993–2006*, Santa Monica, Calif.: RAND Corporation, TR-572-OSD, 2008. As of September 11, 2014: http://www.rand.org/pubs/technical_reports/TR572.html

Gates, Susan M., Edward G. Keating, Bryan Tysinger, Adria Jewell, Lindsay Daugherty, and Ralph Masi, *The Department of the Navy's Civilian Acquisition Workforce: An Analysis of Recent Trends*, Santa Monica, Calif.: RAND Corporation, TR-555-Navy, 2009. As of September 11, 2014: http://www.rand.org/pubs/technical_reports/TR555.html

Gates, Susan M., Elizabeth A. Roth, Sinduja Srinivasan, and Lindsay Daugherty, *Analyses of the Department of Defense Acquisition Workforce*, Santa Monica, Calif.: RAND Corporation, RR-110-OSD, 2013. As of September 11, 2014: http://www.rand.org/pubs/research_reports/RR110.html

Johnston, William, *Civil Service 2000*, Washington, D.C.: The Hudson Institute, 1988.

Kettl, Donald F., Patricia W. Ingraham, Ronald P. Sanders, and Constance Horner, *Civil Service Reform: Building a Government That Works*, Washington, D.C.: The Brookings Institution, 1996.

National Defense Authorization Act for Fiscal Year 1995, Section 342.

U.S. General Accounting Office, *Human Capital: Implementing Pay for Performance at Selected Personnel Demonstration Projects*, Washington, D.C., GAO-04-83, January 2004.

Werber, Laura, Lindsay Daugherty, Edward G. Keating, and Matthew Hoover, *An Assessment of the Civilian Acquisition Workforce, Personnel Demonstration Project*, Santa Monica, Calif.: RAND Corporation, TR-1286, 2012. As of September 11, 2014: http://www.rand.org/pubs/technical_reports/TR1286.html